I0464888

Mostly Planar Motion

John E. Hurtado

Department of Aerospace Engineering

Texas A&M University, College Station

A complete set of notes and examples for a one-semester, sophomore-level dynamics course. Broadly speaking, the content covers point mass and rigid body dynamics in the plane, elementary orbital motions, and elementary rocket dynamics. The principles are presented in a rigorous manner and problems are approached in a systematic way. Furthermore, the notes follow my usual 1-page, 1-topic style.

Preface. This is a thin companion to many sophomore-level dynamics texts. Although this set of notes cannot replace a full-fledged text, its concise form may make it useful for quick reference. My preferences in content and style are reflected in the material and its presentation.

In keeping with a sophomore-level treatment, I've constrained this material in two significant ways. Firstly, only planar motions are covered. (The sole exception is 3-D kinematics using cylindrical coordinates, which could be overlooked.) My reason for keeping things planar is that almost all of the fundamental tools and techniques that are needed to study advanced dynamics can be learned within planar motions. Secondly, there is no mention or use of differential equation techniques to compute solutions to the governing equations. My reason is that the subject of differential equations is beyond the grasp of most sophomore-level students, and it's not needed to understand the evolution of many basic systems.

Certain areas of mathematics are essential to master a subject like dynamics. Particularly, algebra and calculus skills are needed to efficiently and effectively develop and investigate the equations that govern motion. Another essential skill, which is important to engineering in general, is the ability to properly set-up assigned problems; that is, to transform an inquiry into precise mathematical statements. Hopefully, these notes will help a student develop and hone all of these skills.

Many of the examples and exercises are quite similar or identical to problems found in classical texts, and these occurrences are noted. Finally, note that a $\star$ followed by a number (e.g. $\star$ 1) indicates an example or an exercise.

JEH

December 29, 2012

For Juan Sandez and Linda Ann

References and Supplemental Sources. Meriam and Kraige's text is suitable for a sophomore-level dynamics course.

Barger, V. & Olsson M. 1995 *Classical Mechanics: A Modern Perspective.* New York, New York: McGraw-Hill Companies.

Beer, F.P. & Johnston E.R. 2004 *Vector Mechanics for Engineers: Statics and Dynamics, 7ed.* New York, New York: McGraw-Hill Companies.

Greenwood, D.T. 1988 *Principles of Dynamics, 2ed.* Englewood Cliffs, New Jersey: Prentice-Hall.

Meriam, J.L. 1952 *Mechanics, Part II: Dynamics.* John Wiley and Sons, Inc.

Meriam, J.L. & Kraige, L.G. 2007 *Engineering Mechanics: Dynamics, 6ed.* John Wiley and Sons, Inc.

Nelson, E.W., Best, C.L. & McLean, W.G. 1997 *Schaum's Outline of Engineering Mechanics.* New York, New York: McGraw-Hill Companies.

Thomson, W.T. 1986 *Introduction to Space Dynamics.* Mineola, New York: Dover Publications.

Wiesel, W.E. 1989 *Spaceflight Dynamics.* New York, New York: McGraw-Hill Companies.

Contents

4 Point Mass Kinetics in Stationary Frames

7 Point Mass Angular Momentum   120

8 Rigid Body Kinematics   137

9 Apparent Kinematics   163

10 Rigid Body Kinetics   173

1 Preliminaries

An Introduction. This first chapter is used to introduce some of the main concepts that are a part of the principles of motion. These include the differences between kinematics and kinetics, the differences in the types of vectors that are used to model physical entities, the characteristics of reference frames and coordinate systems, and Newton's laws of motion. This material is commonly found in the beginning pages of dynamics textbooks, e.g., chapter 1 of Meriam and Kraige.

The Parts of Engineering Mechanics. [1] Engineering mechanics is typically divided into three branches.

1. Mechanics of point mass models and rigid bodies: this topic deals with external forces and moments acting on the surface or boundary of rigid bodies or point mass models. Point mass models and rigid body models are truly only idealizations.

2. Mechanics of deformable bodies: this topic deals with internal force distributions and deformations that occur when bodies are subject to forces and moments. All real engineering structures are deformable. Sometimes rigid body assumptions are valid or acceptable and sometimes they are not. This material is also commonly called mechanics of materials.

3. Mechanics of fluids: this topic deals with liquids and gases at rest or in motion. More simply, this topic is called fluid mechanics. Fluids can be further modeled as being compressible or incompressible, and this depends on whether the fluid density varies with temperature, pressure, etc.

---

[1] Paraphrased from notes of W.E. Haisler.

Kinematics & Kinetics.   Each one of the branches or areas of engineering mechanics can be further divided into two topics.

1. Kinematics: this topic is concerned with the appearance of motion. The focus is on describing motion without regard to why motion is occuring. Kinematics analysis can provide relationships between position, velocity, and acceleration. Kinematics involves geometry alone and has nothing to do with the laws (Newton's or Euler's) of motion. Therefore, this topic is sometimes referred to as the geometry of motion.

2. Kinetics: this topic deals with bodies subject to forces and moments. Forces and moments can make a body accelerate or not. In the acceleration case it is sometimes said that the forces and moments are *unbalanced*; conversely, in the zero acceleration case it is sometimes said that the forces and moments are *balanced*. Traditionally, the unbalanced case is called the *dynamics problem* whereas the balanced case is called the *statics problem*. The laws of motion (Newton's or Euler's) have everything to do with the kinetics situation.

Different Types of Vectors. Vectors and scalars are central to studying the principles of motion because motion itself, the forces that cause motion, and the object properties that participate in motion can be represented by them. We know what a vector is and that, from a mathematical point of view, magnitude and direction are the important vector quantities. But vectors can represent physical quantities and sometimes the physical quantities have an added importance related to the *location* of the vector. This gives rise to classifications like free vectors, sliding vectors, and bound vectors.

A *free* vector has magnitude and direction but no specified location or point of application. For example, the angular velocity of a rigid body is a free vector. So is a pure torque applied to a rigid body.[2]

A *sliding* vector has magnitude, direction, and a line of action that is aligned with the vector direction. The relevant point is that the affect of the vector is the same regardless of where the vector is placed along the line of action. For example, a force acting on a rigid body is a sliding vector because its influence on the overall motion (translational and rotational) is independent of where the vector is placed along the line of action. (This truth regarding rigid body forces is called the *principle of transmissibility*.)

A *bound* vector has magnitude, direction, and a precise point of application. An applied force on an elastic body is an example of a bound vector.

Operations on vectors are most meaningful when they are explicitly carried out using vector components. But in order to discuss vector components, we must first introduce reference frames and coordinate systems.

---

[2]Truthfully, these free vectors are planar entities. Angular velocity is really angular velocity *in a plane*. And a pure torque is really pure torque *in a plane*. We are able to dress these planar entities as vectors because of the vector-plane equivalency that exists in three dimensions.

Frames of Reference. A frame of reference (or reference frame) is defined as a collection of points such that the distance between the points is constant with respect to time. The minimum number of points required to define a reference frame in three-dimensional space is three noncollinear points. A reference frame is used to make observations regarding motion, but it does not provide a way to *measure* the motion. (For that, we must introduce a coordinate system.)

Reference frames can be inertial or non-inertial. An inertial reference frame is one whose points are either absolutely fixed in space or at most translate relative to an absolutely fixed set of points with the same constant velocity. A non-inertial reference frame is one whose points accelerate with time.

Inertial reference frames are important because it is an axiom of Newtonian mechanics that such frames exist and that his laws of mechanics are valid only in such frames.

Coordinate Systems. A coordinate system can be overlayed on a reference frame. Indeed, a reference frame can be home to an infinite number of coordinate systems, but typically one associates one reference frame to one coordinate system and the two become synonymous.

A coordinate system is a group of objects within a reference frame that allows measurement. In three-dimensional space, the system is constructed by three mutually orthogonal unit vectors in a reference frame, called the axes of the coordinate system or the coordinate axes or the reference vectors, that meet at a point in a reference frame, called the origin.[3] The set of unit vectors always form a right-handed system. As an example, we could define a coordinate system a^+ to have coordinate axes a_i meeting at an origin o in some reference frame.[4]

$$a^+ : \{o,\ \hat{a}_1,\ \hat{a}_2,\ \hat{a}_3\} \quad \text{or} \quad a^+ : \{o,\ \hat{a}_i\}, \quad i = 1, 2, 3 \tag{7.1}$$

We would call this coordinate system a^+ in the reference frame, or because of the close relationship between reference frames and coordinate systems, simply the a^+ frame or frame a^+. Furthermore, the coordinate axes $\hat{a}_i$ will often be simply called the a^+ axes.

Measurement in a reference frame is done relative to the coordinate system (or frame) origin and along the coordinate (or frame) axes. Thus, the location of an arbitrary point in three-dimensional space can be described by the elements of the coordinate system. All that is needed are measures along the coordinate axes.

$$p = p_1\hat{a}_1 + p_2\hat{a}_2 + p_3\hat{a}_3 \tag{7.2}$$

Here, the p_i are scalar coefficients of the vector p in the a^+ frame. These scalar coefficients are commonly called the *components* of p or the *measures* of p.

---

[3] Strictly speaking, the axes don't have to meet at a point.
[4] Right-handed means, for example, $\hat{a}_1 \times \hat{a}_2 = \hat{a}_3$. Orthogonal means $\hat{a}_i \cdot \hat{a}_j = 0$ if $i \neq j$. And unit vector means $\|\hat{a}_i\| = 1$.

Newton's Laws. Newton's original laws of motion may be stated using modern language in the following way.

N1L — A point mass remains at rest or continues to move with uniform velocity, which means with constant speed along a straight line, if the forces acting on it are balanced.

N2L — A point mass accelerates in an amount proportional to, and in the direction of, the unbalance of forces acting on it.

N3L — The forces of action and reaction between contacting point masses have equal magnitudes and opposite, collinear directions.

N4L — A gravitational law governs the mutual attractive force between point masses as described by the limerick:

If ρ be the distance to O

Sir Newton said he could show

That the force of attraction

Behaves like the fraction

Of one over the square of ρ

(R.M. Rosenberg)

Newton's second law (N2L) gives us $f = ma$ where a is the acceleration in a non accelerating reference frame. Newton's fourth law (N4L) gives us $f = Gm_1m_2/\rho^2$ where G is a universal constant. N2L and N4L combine to give that the amount of gravity at the Earth's surface is $g_0 = Gm_e/R^2 = 9.825 \text{ m/s}^2$. At an altitude h above the Earth's surface, the gravity behaves according to $g = g_0 R^2/(R+h)^2$.

2 Point Mass Kinematics in Stationary Frames

An Introduction. The purpose of this chapter is to refresh our understanding of motion along fixed directions.

One important fact that often gets overlooked is that *inertial kinematic vectors* are critical when studying dynamics. Inertial kinematic vectors encompass inertial position vectors, inertial velocity vectors, and inertial acceleration vectors. An inertial position vector is one that is measured from a fixed point.[5] An inertial velocity vector is the time derivative of this inertial position vector as viewed by an inertial observer, i.e., an observer in an inertial reference frame. An inertial acceleration vector is the time derivative of the an inertial velocity vector as viewed by an inertial observer.

In this section, the kinematic vectors will be measured along fixed directions, or measured in fixed reference frames. This doesn't have to be so, but it's what we'll do for now.

---

[5] Truly, the point can move with constant velocity along a straight line.

Rectilinear Motion. This begins a kinematic analysis of rectilinear motion, including position vectors, velocity vectors, and acceleration vectors.

Rectilinear motion means motion along a fixed, straight line. The line can be defined by an axis (e.g., the $\hat{n}_1$ axis) of an orthogonal reference frame whose origin is located at a fixed point o. The instantaneous position of a point along the axis is measured by the coordinate x.

Commonly, information about the instantaneous position (x) or velocity (v) or acceleration (a) is given, but information about the instantaneous position or velocity or acceleration is sought. There are five common situations to consider.

1. $x(t)$ is known: the magnitude of the instantaneous velocity and acceleration can be determined from differentiation.

2. $v(t)$ is known: the magnitude of the instantaneous position and acceleration can be determined from integration and differentiation, respectively.

3. $a(t)$ is known: the magnitude of the instantaneous position and velocity can be determined from integration.

4. $a(x)$ is known: beginning with $a = dv/dt$, the chain rule of calculus gives $v\,dv = a(x)\,dx$, which can be integrated to give $v(x)$. Integrating $dx/v(x) = dt$ gives an expression that relates time to position $t(x)$, which can be inverted to determine position as a function of time, $x(t)$.

5. $a(v)$ is known: beginning with $a = dv/dt$, we can find an expression that relates time to velocity. Once $t(v)$ is formally known, the expression may be inverted to give velocity as a function of time, $v(t)$. The position as a function of time immediately follows.

★ 1 Descent of a Lunar Lander.

Consider the descent of a lunar lander.
Suppose it experiences an acceleration equal to
$$a = 0.6t - g$$
where g is a constant. Let $g = 0.2 \frac{m}{s^2}$
Find expressions that relate
velocity & position to time.

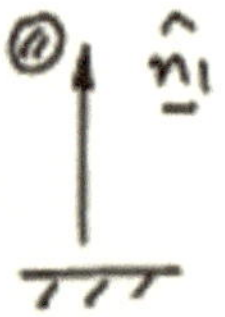

Note $\underline{p} = x \, \hat{n_1}$ $\underline{v} = v \, \hat{n_1}$

$$\underline{g} = a \, \hat{n_1}$$

Also $v = dx/dt$ & $a = dv/dt$

Integrating

$$dv = (0.6t - g) dt \qquad v = 0.6\frac{t^2}{2} - gt + v_0$$

And similarly

$$x = 0.6\frac{t^3}{6} - g\frac{t^2}{2} + v_0 t + x_0$$

Suppose one desires $v(10) = 0$ and $x(10) = 0$.
Then compute the corresponding initial conditions.

$$v(0) = -10 \, m/s \qquad and \qquad x(0) = 100 \, m$$

★ 2 Escape Velocity.

Newton's inverse square law means that the acceleration of a point mass depends on its distance away from the Earth, $a = -gR^2/r^2$, where g is the amount of gravity at the surface.

The motion of the point mass is rectilinear if one considers motion along the radial direction. Let's compute the escape velocity, which is the speed the point mass must have at the Earth's surface to escape.

$$\underline{a} = a\,\hat{\underline{n}} = -gR^2/r^2\,\hat{\underline{n}}$$

Using the chain rule

$$a = \frac{dv}{dt} = \frac{dv}{dr}\frac{dr}{dt} = \frac{dv}{dr}v$$

This gives

$$\int_R^r -\frac{gR^2}{r^2}\,dr = \int_{v_0}^{v} v\,dv$$

$$\frac{gR^2}{r} - gR = \frac{1}{2}\left(v^2 - v_0^2\right)$$

The escape velocity conditions means $v = 0$ when $r = \infty$. Thus

$$v_0 = \sqrt{2gR}$$

⋆ 3 Escape Altitude.

One can compute the altitude as a function of time for Newton's inverse square model from the previous page.

Note, if $v_0 = \sqrt{2gR}$, then $v = \sqrt{2gR^2/r}$

Integrating

$$\int_R^r \frac{dr}{\sqrt{\dfrac{2gR^2}{r}}} = \int_0^t dt$$

$$t(r) = \frac{1}{3R}\sqrt{\frac{2}{g}}\left(r^{3/2} - R^{3/2}\right)$$

or

$$r(t) = \left\{3Rt\sqrt{9/2} + R^{3/2}\right\}^{2/3}$$

Constant Acceleration & Distance Traveled. Rectilinear motion involving constant acceleration is a practical, special case to consider. A familiar example of this is the vertical motion of point mass in an idealized constant gravity field, $a = -9.81$ m/s^2. The previous results with a equal to a known constant c are useful to find explicit expressions relating the instantaneous velocity, position, and time.

$$v = ct + v_0 \tag{15.1}$$

$$x = \frac{c}{2}t^2 + v_0 t + x_0 \tag{15.2}$$

$$v^2 = v_0^2 + 2c(x - x_0) \tag{15.3}$$

Note that the initial time t_0 is assumed to be zero in the above equations, and that x_0 and v_0 are initial conditions.

When the acceleration is constant, these relationships are really all there is to know: (15.1) gives the instantaneous velocity as a function of time, or time as a function of velocity; (15.2) gives the instantaneous position as a function of time, or time as a function of position; and (15.3) gives the instantaneous velocity as a function of instantaneous position, or position as a function of velocity.

The distance traveled by a point mass can be computed from a modified version of $x(t) = \int_{t_0}^{t} v(s)\, ds + x_0$.

$$d(t) = \left| \int_{t_0}^{t_1} v(s)\, ds \right| + \left| \int_{t_1}^{t_2} v(s)\, ds \right| + \ldots + \left| \int_{t_{n-1}}^{t} v(s)\, ds \right| \tag{15.4}$$

Here, like before, t_0 is the initial time. The intermediate times between t_0 and t are instances when the velocity magnitude undergoes a sign change. These times must be identified if they are not given.

Planar Motion. Point mass kinematics in the plane can be studied using Cartesian coordinates.

The instantaneous position of a point on a plane needs two independent coordinates to pinpoint its location.

$$p = x\,\hat{n}_1 + y\,\hat{n}_2 \qquad\qquad (16.1)$$

The measures x and y can vary, but the axes $\hat{n}_1$ and $\hat{n}_2$ are fixed: they are unit vectors that do not change direction.

The time derivative of the instantaneous position vector gives the instantaneous velocity vector, and a subsequent time derivative gives the instantaneous acceleration vector.

$$v = \dot{x}\,\hat{n}_1 + \dot{y}\,\hat{n}_2 = v_x\,\hat{n}_1 + v_y\,\hat{n}_2$$
$$a = \ddot{x}\,\hat{n}_1 + \ddot{y}\,\hat{n}_2 = \dot{v}_x\,\hat{n}_1 + \dot{v}_y\,\hat{n}_2 = a_x\,\hat{n}_1 + a_y\,\hat{n}_2$$

$$(16.2, 16.3)$$

The overdot means the time derivative of the scalar measure, e.g., $\dot{x} \equiv dx/dt$.

It is important to note that motion appears uncoupled using this *Cartesian coordinate* description. That is, we could have full knowledge of x, v_x, or a_x along the $\hat{n}_1$ direction and determine through differentiation or integration the other unknown kinematic variables along the $\hat{n}_1$ direction. Likewise for the kinematics along the $\hat{n}_2$ direction.

⋆ 4 A Fired Missile.

A missile is fired in the horizontal direction from an aircraft at an altitude of 981 m traveling at 900 km/hr = 250 m/s. The missile is subject to a horizontal thrust resulting in a horizontal acceleration of 0.6 g. Compute the line-of-sight angle between the launch point and impact point.

$$\underline{P_x} = x\ \underline{\hat{n}_1}$$

$$\underline{P_y} = y\ \underline{\hat{n}_2}$$

$$y(t) = g\ \frac{t^2}{2}$$

$$981 = 9.81\ \frac{t^2}{2}. \quad \text{Solving gives } t = 14.14 \text{ s}$$

$$x(t) = a_x\ \frac{t^2}{2} + v_{0x}\ t$$

$$x(t) = (0.6)9.81\ \frac{t^2}{2} + 250\ t$$

Solving gives $x(14.14) \approx 4124$ m

So $\tan\theta = \dfrac{981}{4124}$ or $\theta = 13.38\,^\circ$

Projectile Motion. Projectile motion is idealized motion of a point mass near the Earth's surface. In a Cartesian coordinate description, the acceleration component in the horizontal $\hat{n}_1$ direction is zero whereas the acceleration component in the vertical $\hat{n}_2$ direction is the negative of the gravitational constant. Consequently, the velocity component along the $\hat{n}_1$ direction is constant and the position component along the $\hat{n}_1$ direction is a linear function of time. Motion along $\hat{n}_2$ is governed by the familiar constant acceleration formulae.

Horizontal	Vertical
$a_x = 0$	$v_y^2 = v_{y0}^2 - 2g(y - y_0)$
$v_x = v_{x0}$	$v_y = -gt + v_{y0}$
$x = v_{x0}t + x_0$	$y = -\frac{1}{2}gt^2 + v_{y0}t + y_0$

$$(18.1)$$

Often, the components of velocity are written using the launch speed and launch angle measured from the horizon.

$$v_{x0} = v\cos\gamma \quad ; \quad v_{y0} = v\sin\gamma \qquad (18.2, 18.3)$$

These equations can be manipulated to show a few properties of projectile motion over level terrain:

1. The range (i.e., the x component of position) is maximized for a 45 degree launch angle;

2. The time of flight is maximized for a 90 degree launch angle.

⋆ 5 Extremes in Projectile Motion.

The equations for x & y motions of a projectile can be manipulated to investigate the maximum range & maximum time-of-flight.

For maximum range : $x = v_0 \cos\gamma\, t$ or

$$t = x / v_0 \cos\gamma$$

This leads to

$$y(x) = -\frac{1}{2} g \frac{x^2}{v_0^2 \cos^2\gamma} + x \frac{\sin\gamma}{\cos\gamma}$$

Setting $y(x) = 0$ gives $0 = x \left(\sin\gamma \cos\gamma - \frac{gx}{2v_0^2}\right)$

or $x = \frac{v_0^2}{g} \sin 2\gamma$

Setting $dx/d\gamma$ to zero

$$0 = \frac{2 v_0^2}{g} \cos 2\gamma$$ or $\gamma = 45°$

For maximum time-of-flight :

$$y = -\frac{1}{2} g t^2 + t\, v_0 \sin\gamma$$

Setting y to zero gives $0 = t\left(v_0 \sin\gamma - \frac{1}{2} gt\right)$
or

$$t = \frac{2 v_0 \sin\gamma}{g}$$

Setting $dt/d\gamma$ to zero gives $\gamma = 90°$

The Trajectory Space. The admissible trajectory space of projectile motion in an unobstructed region are those (x, y) points that can be reached.

Given a launch velocity v, the admissible trajectory space is governed by a quadratic equation in the tangent of the launch angle.

$$\tan^2 \gamma - \frac{2v^2}{gx} \tan\gamma + \left(\frac{2v^2 y}{gx^2} + 1 \right) = 0 \qquad (20.1)$$

The explicit solutions to eq. (20.1) reveal the different scenarios.

$$\tan\gamma = \frac{v^2}{gx} \pm \sqrt{\frac{v^4}{g^2 x^2} - \frac{2v^2 y}{gx^2} - 1} \qquad (20.2)$$

These solutions are telling: there are two launch angles for all (x, y) points strictly *inside* the trajectory space, and for this case the radicand is positive; there is only one launch angle for all (x, y) points strictly *on the edge of* the trajectory space (called the trajectory envelope), and for this case the radicand is zero; and there are nonsensical launch angles for all (x, y) points strictly *outside* the trajectory space, and in this case the radicand is negative.

The *trajectory envelope* is defined when the radicand equals zero. These are the points that can barely be reached. Notice then, that the maximum height of the trajectory envelope is found from setting $x = 0$, and the maximum range of the trajectory envelope is found from setting $y = 0$.

$$\text{max height } y = \frac{v^2}{2g} \quad ; \quad \text{max range } x = \frac{v^2}{g} \qquad (20.3, 20.4)$$

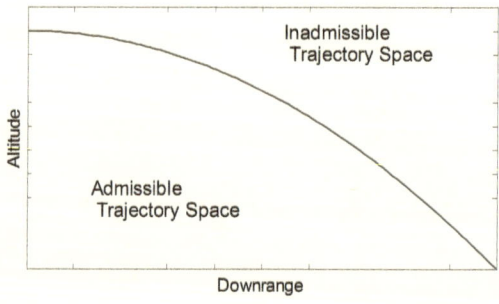

Chapter 2 Problem Set.

1. (M&K 2.13) The catapult that launches a jet aboard a stationary aircraft carrier provides a constant acceleration along a straight path. (a) Write kinematic equations for the velocity and position of the jet as it travels along the launch runway. (b) Compute the constant acceleration (in g's) if the catapult provides a launch velocity of 200 mi/hr in a distance of 330 ft for a jet beginning from rest. Answer: $a = 4.05$ g.

2. (M&K 2.18) A lunar landing module uses its engine and thrusters to descend to a target altitude above the lunar surface with a safe downward velocity. After reaching the target altitude, the engine and thrusters are abruptly cut off, and the landing module falls along a straight vertical path with a constant lunar surface gravity which is equal to 1/6 the gravity at the Earth's surface. (a) Write kinematic equations for the velocity and position of the landing module after the engine and thrusters cut off. (b) Compute the impact velocity of the landing module from a target altitude of 5 m and downward velocity of 2.5 m/s. Answer: $v = 4.75$ m/sec.

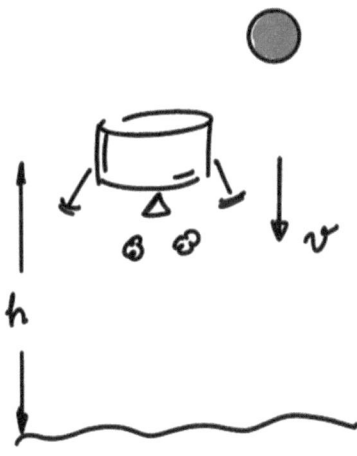

3. (M&K 2.43) Point mass models that are studied close to a heavenly body's surface are assumed to experience a constant gravity g_0. Point mass models that are studied far away from a heavenly body's surface are assumed to experience a gravity that changes depending on the distance from the body's center given by $g = g_0 R^2/(R+h)^2$, where R is the radius of the heavenly body and h is the altitude of the point mass above the body's surface. The difference in computed speeds using these different models can be significant. Consider a chunk of orbital debris that is instantaneously at rest high above the lunar surface. The diameter of the Moon is 2160 miles and the gravity at the lunar surface is g_0 =5.37 ft/s². (a) Write a kinematic equation for the velocity of a point mass assuming a constant gravity acceleration, and compute the impact speed if the debris begins at an altitude of 750 miles. (b) Develop or write a kinematic equation for the velocity of a point mass assuming a gravity acceleration that varies according to $g = g_0 R^2/(R+h)^2$, and compute the impact speed if the debris begins at an altitude of 750 miles. Answers: $v = 6522$ ft/sec and $v = 5010$ ft/sec.

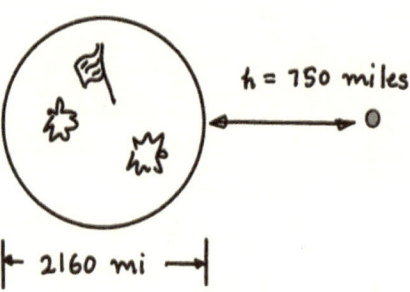

4. (M&K 2.56) A tall building has N stories and each story measures h ft in height. An experimental ball is released from rest beginning at the Nth story. Gravity is constant and acts along the vertical direction. The motion of the ball is only along the vertical direction. (a) Write kinematic equations for the velocity and position of the ball. (b) Show that an expression for the time it takes for the ball to pass between any two floors is given by $\Delta t = \sqrt{2h/g}\,(\sqrt{n} - \sqrt{n-1})$, where h is the height of each story, and n designates a particular story when counted from the top. (c) Let $\sqrt{2h/g} = 1$, and plot Δt as a function of n from $n = 1$ to 100. Some examples include the following for a building with each story measuring 16.1 ft in height:

- From the roof top to the first floor from the top, $\Delta t = 1$ sec
- From the $N - 9$th story from the top to the $N - 10$th story from the top, $\Delta t = 0.1623$ sec ;

Hints. Compute the time to arrive at a particular story, and then difference the appropriate times.

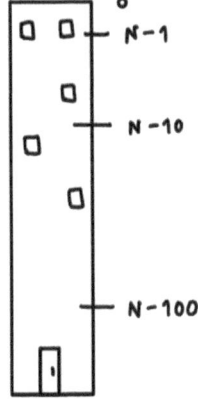

5. Sometimes an acceleration profile abruptly changes and the kinematics must be pieced together. Consider a point mass that undergoes the acceleration history shown in the figure. During the first two seconds, the acceleration is a linear function of time. Thereafter, the acceleration is constant. The initial position of the point mass is $x(0) = 4/3$ m and the initial velocity is $v(0) = -1$ m/s. (a) Develop the velocity and position kinematic equations that are valid during the first two seconds, and compute the velocity and position at the two second mark. (b) Develop the velocity and position kinematic equations that are valid during the interval from two seconds to four seconds, and compute the velocity and position at the four second mark. Answer: $x = 12$ m.

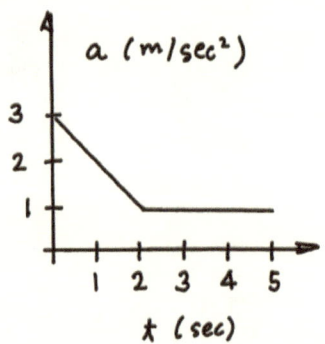

6. (Mer 644) For speeds up to a certain limit, the resistance to rectilinear motion in air or another fluid can be modeled as kv^n where $k > 0$ and $n > 1$ are constants. The resistance is essentially a deceleration. Find an expression for the instantaneous speed as a function of time, initial speed v_0, and constants k and n.

7. (B&J 11.111) A model rocket is launched from point o with an initial velocity v_0 equal to 90 m/s. Unfortunately, the rocket has no parachute and crash lands 105 meters downrange from point o. Gravity is constant and acts along the negative vertical direction. (a) Write the kinematic equations that govern the planar motion of the rocket. (b) Compute the launch angle (measured from the horizontal direction). (c) Compute the maximum height attained by the rocket. (d) Compute the time of flight. (e) Does the rocket touch its trajectory envelope? Why or why not? Answer: $\gamma = 86.3$ deg.

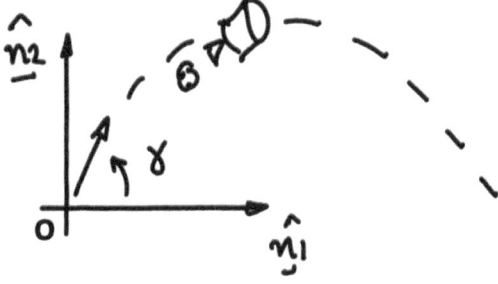

8. Consider projectile motion over a flat surface. Let the launch velocity equal 100 m/s. For simplicity, let gravity equal 10 m/s². Perform the following:

a. Show that points on the trajectory envelope are governed by a quadratic equation.

$$2y = v^2/g - gx^2/v^2$$

b. Calculate and plot 10 points that lie on the trajectory envelope.

c. Compute the launch angles for the 10 points in part b.

d. Compute the launch angles for 3 points the lie within the trajectory envelope.

e. Perform the mathematical manipulations to show that an expression for the time to hit the trajectory envelope is $t_e = v/(g \sin \gamma)$.

f. Perform the mathematical manipulations to show that an expression for the ratio of time to hit the trajectory envelope and time of flight is $t_e/t_f = 1/(2 \sin^2 \gamma)$.

g. Perform the mathematical manipulations to show that an expression for the ratio of time to hit the trajectory envelope and time to reach the maximum height is $t_e/t_{max} = 1/\sin^2 \gamma$.

9. Consider a launch velocity of $v = 100$ m/s. For simplicity, let $g = 10$ m/s^2. Plot the (x, y) projectile motion for the following six launch angles in the table. Complete the table by computing the time to hit the trajectory envelope and the time of flight for each launch angle. Plot the trajectory envelope. Overall, your graphs should look similar to what's shown below.

Launch angle (deg)	Time envelope (sec)	Time of flight (sec)
78.1113		
64.6538		
53.6156		
45.0000		
30.0000		
15.0000		

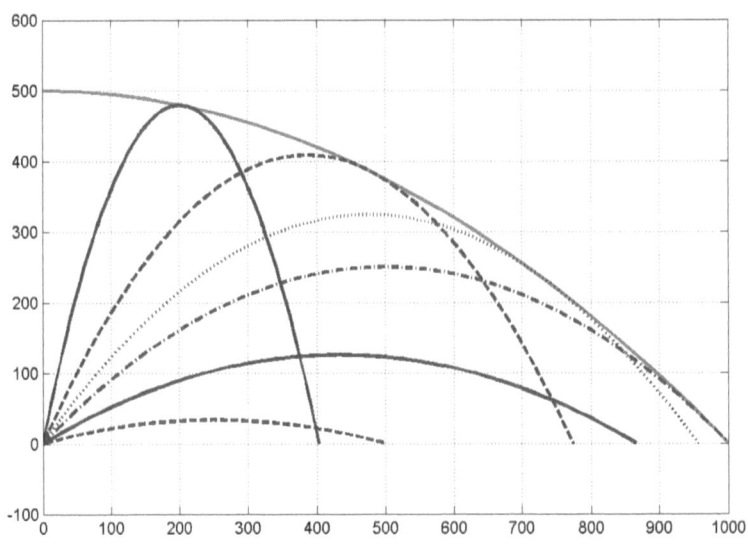

10. (M&K 2.100) Show that the launch angle γ that will maximize the down range distance along an incline with angle β is given by the linear expression $\gamma = \beta/2 + \pi/4$.

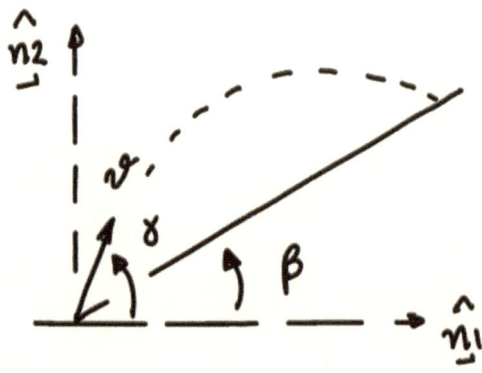

Chapter 2 Notes.

3 Point Mass Kinematics using Rotating Frames

An Introduction. Although an inertial position vector is measured *from* a fixed point, it does not have to be measured *along* fixed directions. That is, an inertial position vector can be measured from a fixed point but along directions that change. This begins a study of kinematics variables expressed in rotating reference frames.

Here's where things can get confusing: an inertial kinematic vector is measured *from* a fixed point; it doesn't have to measured *along* fixed directions; but its time derivatives must be as *seen* by a fixed observer. These three concepts are covered on the next page.

Rotating Reference Frames. When describing the planar motion of a point mass, it is occasionally convenient to express the vector kinematics in a non inertial, rotating reference frame. Consider the inertial location of a point in a plane.

$$\boldsymbol{p} = x\,\hat{\boldsymbol{n}}_1 + y\,\hat{\boldsymbol{n}}_2 = r\hat{\boldsymbol{e}}_1 \tag{32.1}$$

These are two valid expressions for the location of a point. The first uses Cartesian coordinates along unmoving $\hat{\boldsymbol{n}}_1$ and $\hat{\boldsymbol{n}}_2$ axes, whereas the second uses a polar coordinate description. Here, $\hat{\boldsymbol{e}}_1$ is a unit vector that always points from a fixed origin o to the instantaneous location of the moving point. Consequently, the $\hat{\boldsymbol{e}}_1$ unit vector changes: its magnitude stays the same, but its direction changes.

The instantaneous inertial velocity vector is still the time derivative of the instantaneous inertial position vector as seen by a fixed observer.

$$\boldsymbol{v} = \frac{\mathrm{d}}{\mathrm{d}t}\,(\boldsymbol{p}) = \frac{\mathrm{d}}{\mathrm{d}t}\,(r\hat{\boldsymbol{e}}_1) = \dot{r}\hat{\boldsymbol{e}}_1 + r\frac{\mathrm{d}}{\mathrm{d}t}(\hat{\boldsymbol{e}}_1) \tag{32.2}$$

The time derivative of $\hat{\boldsymbol{e}}_1$ as seen by a fixed observer must be determined, and this is most easily done by noting that $\hat{\boldsymbol{e}}_1$ can be expressed along the $\hat{\boldsymbol{n}}_1$ and $\hat{\boldsymbol{n}}_2$ directions.

$$\hat{\boldsymbol{e}}_1 = \cos\theta\,\hat{\boldsymbol{n}}_1 + \sin\theta\,\hat{\boldsymbol{n}}_2 \tag{32.3}$$

Now, taking a time derivative of this expression shows that the change in $\hat{\boldsymbol{e}}_1$ is related to the angular orientation of $\hat{\boldsymbol{e}}_1$ and a new unit vector that is orthogonal to $\hat{\boldsymbol{e}}_1$.

$$\frac{\mathrm{d}}{\mathrm{d}t}(\hat{\boldsymbol{e}}_1) = \dot{\theta}\hat{\boldsymbol{e}}_2 \tag{32.4}$$

This result can be used to rewrite the velocity vector.

$$\boldsymbol{v} = \dot{r}\hat{\boldsymbol{e}}_1 + r\dot{\theta}\hat{\boldsymbol{e}}_2 = v_r\hat{\boldsymbol{e}}_1 + v_\theta\hat{\boldsymbol{e}}_2 \tag{32.5}$$

This simple example illustrated that we measured the inertial position vector *from* a fixed point, but *along* non-fixed directions, and computed the time derivative as *seen* by a fixed observer.

Relating One Set of Unit Vectors to Another. Consider unit vectors $\hat{\boldsymbol{b}}_1$, $\hat{\boldsymbol{b}}_2$ that appear rotated relative to unit vectors $\hat{\boldsymbol{n}}_1$, $\hat{\boldsymbol{n}}_2$. It is often convenient (see the previous page!) to express the $\hat{\boldsymbol{b}}$ unit vectors in terms of the $\hat{\boldsymbol{n}}$ unit vectors and vice versa.

$$
\begin{aligned}
\hat{\boldsymbol{b}}_1 &= \cos\theta\,\hat{\boldsymbol{n}}_1 + \sin\theta\,\hat{\boldsymbol{n}}_2 \\
\hat{\boldsymbol{b}}_2 &= -\sin\theta\,\hat{\boldsymbol{n}}_1 + \cos\theta\,\hat{\boldsymbol{n}}_2
\end{aligned}
\tag{33.1}
$$

$$
\begin{aligned}
\hat{\boldsymbol{n}}_1 &= \cos\theta\,\hat{\boldsymbol{b}}_1 - \sin\theta\,\hat{\boldsymbol{b}}_2 \\
\hat{\boldsymbol{n}}_2 &= \sin\theta\,\hat{\boldsymbol{b}}_1 + \cos\theta\,\hat{\boldsymbol{b}}_2
\end{aligned}
\tag{33.2}
$$

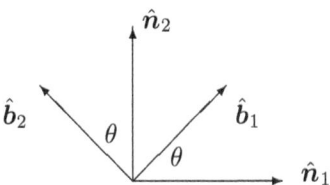

⋆ 6 A Two-Link Robot Arm.

A two-link robot arm moves
in a horizontal plane. Each
link has length L. Find
expressions for the position
of the end relative to the
base.

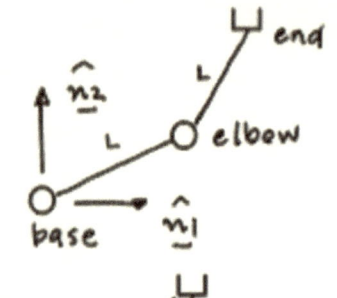

To begin $\underline{p} = \underline{p}_1 + \underline{p}_2$

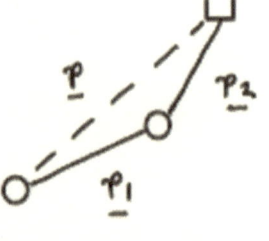

Next introduce rotating
reference frames e^+ &
b^+. Note

$$\underline{p}_1 = L \; \hat{\underline{e}}_1 \; ; \; \underline{p}_2 = L \; \hat{\underline{b}}_1$$

So

$$\underline{p} = L \; \hat{\underline{e}}_1 + L \; \hat{\underline{b}}_1$$

We can write $\underline{p}$ solely
in the e^+ frame

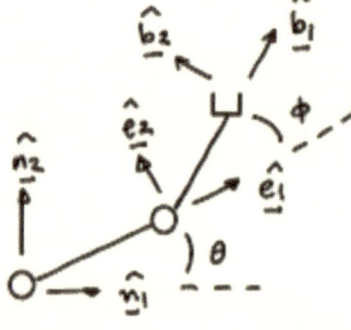

$$\underline{p} = L \; \hat{\underline{e}}_1 + L \; (\cos\phi \; \hat{\underline{e}}_1 + \sin\phi \; \hat{\underline{e}}_2)$$
$$= (L + L \cos\phi) \; \hat{\underline{e}}_1 + L \sin\phi \; \hat{\underline{e}}_2$$

Or we can write $\underline{p}$ solely in the b^+ frame

$$\underline{p} = L \; (\cos\phi \; \hat{\underline{b}}_1 - \sin\phi \; \hat{\underline{b}}_2) + L \; \hat{\underline{b}}_1$$
$$= (L + L \cos\phi) \; \hat{\underline{b}}_1 - L \sin\phi \; \hat{\underline{b}}_2$$

Planar Kinematics in Polar Coordinates. The velocity vector in terms of polar coordinates and rotating unit vectors was determined, $v = \dot{r}\hat{e}_1 + r\dot{\theta}\hat{e}_2 = v_r\hat{e}_1 + v_\theta\hat{e}_2$. The acceleration vector can be determined in a similar way.

$$a = (\ddot{r} - r\dot{\theta}^2)\hat{e}_1 + (r\ddot{\theta} + 2\dot{r}\dot{\theta})\hat{e}_2 = a_r\hat{e}_1 + a_\theta\hat{e}_2 \tag{35.1}$$

As an exercise, you should derive equation (35.1). You will encounter the time derivative of $\hat{e}_2$, which is $d\hat{e}_2/dt = -\dot{\theta}\hat{e}_1$.

These new kinematic equations for position, velocity, and acceleration are occasionally called the *polar coordinate* description.

Cartesian

$p = x\,\hat{n}_1 + y\,\hat{n}_2$

$v = \dot{x}\hat{n}_1 + \dot{y}\hat{n}_2$

$a = \ddot{x}\hat{n}_1 + \ddot{y}\hat{n}_2$

Polar

$p = r\,\hat{e}_1$

$v = \dot{r}\,\hat{e}_1 + r\dot{\theta}\,\hat{e}_2$

$a = (\ddot{r} - r\dot{\theta}^2)\hat{e}_1 + (r\ddot{\theta} + 2\dot{r}\dot{\theta})\hat{e}_2$

$$\tag{35.2}$$

The polar coordinate description is convenient when a device is able to measure range to target (r and its time derivatives) and elevation above the horizon (θ and its time derivatives).

⋆ 7 Radar Tracking of a Rocket.

A rocket undergoes vertical flight with speed v and acceleration a . A radar tracking system a distance b away from the launch site is used to record the motion. Find the speed v in terms of b, θ, and $\dot{\theta}$.

The position & velocity vectors can be written.

$$\underline{p} = b\,\hat{\underline{n}}_1 + h\,\hat{\underline{n}}_2$$

$$\underline{v} = \dot{h}\,\hat{\underline{n}}_2 = v\,\hat{\underline{n}}_2$$

But $\underline{p} = r\,\hat{\underline{e}}_1$

and

$$\underline{v} = \dot{r}\,\hat{\underline{e}}_1 + r\dot{\theta}\,\hat{\underline{e}}_2$$

Also

$$\hat{\underline{n}}_2 = \cos\theta\,\hat{\underline{e}}_2 + \sin\theta\,\hat{\underline{e}}_1$$

So $\underline{v} = v\cos\theta\,\hat{\underline{e}}_2 + v\sin\theta\,\hat{\underline{e}}_1$

This gives $v\sin\theta = \dot{r}$ and $v\cos\theta = r\dot{\theta}$ which gives

$$v = \frac{b\dot{\theta}}{\cos^2\theta}$$

One can compute the acceleration, too.

$$a = \frac{b}{\cos^2\theta}\left(\ddot{\theta} + 2\dot{\theta}^2\tan\theta\right)$$

Normal and Tangential Velocity Coordinates. The polar coordinate description for planar kinematics was motivated by the desire to express the inertial position vector using one measurement along a changing unit vector direction.

$$\boldsymbol{p} = r\hat{\boldsymbol{e}}_1 \qquad \text{instead of} \qquad \boldsymbol{p} = x\,\hat{\boldsymbol{n}}_1 + y\,\hat{\boldsymbol{n}}_2 \tag{37.1}$$

We could ask that something similar be done for the velocity vector.

$$\boldsymbol{v} = v\hat{\boldsymbol{e}}_t \qquad \text{instead of} \qquad \boldsymbol{v} = \dot{x}\,\hat{\boldsymbol{n}}_1 + \dot{y}\,\hat{\boldsymbol{n}}_2 = \dot{r}\,\hat{\boldsymbol{e}}_1 + r\dot{\theta}\,\hat{\boldsymbol{e}}_2 \tag{37.2}$$

The new direction is labeled $\hat{\boldsymbol{e}}_t$; the subscript is meant to indicate "tangential" because the velocity vector is tangent to the path. (This is similar to using the moniker $\hat{\boldsymbol{e}}_r$ in polar coordinates, like some folks do.) Like the unit vectors $\hat{\boldsymbol{e}}_1$ and $\hat{\boldsymbol{e}}_2$ that are part of the polar coordinate description, this new direction unit vector is non inertial. It is part of a new rotating reference frame. The unit vector $\hat{\boldsymbol{e}}_t$ is oriented with respect to the $\hat{\boldsymbol{n}}_1$ axis through a new angle β.

The instantaneous acceleration vector can be determined through vector differentiation, where one will encounter the time derivative of $\hat{\boldsymbol{e}}_t$.

$$\boldsymbol{a} = \dot{v}\,\hat{\boldsymbol{e}}_t + v\frac{\mathrm{d}}{\mathrm{d}t}\hat{\boldsymbol{e}}_t = \dot{v}\,\hat{\boldsymbol{e}}_t + v\dot{\beta}\,\hat{\boldsymbol{e}}_n = \dot{v}\,\hat{\boldsymbol{e}}_t + \frac{v^2}{\rho}\,\hat{\boldsymbol{e}}_n \tag{37.3}$$

The last expression uses a relationship between the time derivative of β and the *instantaneous radius of curvature*, ρ: $v = \rho\dot{\beta}$. In general, ρ changes, and this treatment, unlike common presentations of this same material, lets ρ be positive or negative.

Equations (37.2) and (37.3) are expressions for instantaneous velocity and acceleration using normal and tangential coordinates.

These expressions are most useful when the path of the particle is already known, especially for known paths with a constant ρ.

A Collection of Kinematic Coordinates. Some various forms of kinematic expressions for general planar motion are gathered here.

General Planar Motion		
Cartesian	Polar	Norm. & Tang.
(x, y)	(r, θ)	(v, ρ)
$\boldsymbol{p} = x\,\hat{\boldsymbol{n}}_1 + y\,\hat{\boldsymbol{n}}_2$	$\boldsymbol{p} = r\,\hat{\boldsymbol{e}}_1$	
$\boldsymbol{v} = \dot{x}\hat{\boldsymbol{n}}_1 + \dot{y}\hat{\boldsymbol{n}}_2$	$\boldsymbol{v} = \dot{r}\,\hat{\boldsymbol{e}}_1 + r\dot{\theta}\,\hat{\boldsymbol{e}}_2$	$\boldsymbol{v} = v\hat{\boldsymbol{e}}_t$
$\boldsymbol{a} = \ddot{x}\hat{\boldsymbol{n}}_1 + \ddot{y}\hat{\boldsymbol{n}}_2$	$\boldsymbol{a} = (\ddot{r} - r\dot{\theta}^2)\hat{\boldsymbol{e}}_1 + (r\ddot{\theta} + 2\dot{r}\dot{\theta})\hat{\boldsymbol{e}}_2$	$\boldsymbol{a} = \dot{v}\hat{\boldsymbol{e}}_t + v^2/\rho\,\hat{\boldsymbol{e}}_n$

The kinematic equations can be simplified for the special case of a point mass undergoing circular motion. In the polar coordinate description $\dot{r}$ is set zero, and in the normal & tangential coordinate description ρ is constant and equals r. Moreover, the angle β of the normal and tangential velocity coordinate description is related to the angle θ of the polar coordinate description, $\beta = \theta + \pi/2$. Thus, $\hat{\boldsymbol{e}}_2$ and $\hat{\boldsymbol{e}}_t$ are collinear and point in the same direction, whereas $\hat{\boldsymbol{e}}_1$ and $\hat{\boldsymbol{e}}_n$ are collinear and point in opposite directions.

Circular Motion	
Polar	Norm. & Tang.
(r, θ)	(v, r)
$\boldsymbol{p} = r\,\hat{\boldsymbol{e}}_1$	
$\boldsymbol{v} = r\dot{\theta}\,\hat{\boldsymbol{e}}_2$	$\boldsymbol{v} = v\hat{\boldsymbol{e}}_t$
$\boldsymbol{a} = -r\dot{\theta}^2\hat{\boldsymbol{e}}_1 + r\ddot{\theta}\hat{\boldsymbol{e}}_2$	$\boldsymbol{a} = v^2/r\,\hat{\boldsymbol{e}}_n + \dot{v}\,\hat{\boldsymbol{e}}_t$

Satellite Speed in a Circular Orbit.   If a small body is in a constant speed, circular orbit around a larger body whose location is fixed in space, then the kinematic expressions using normal and tangential components are $\boldsymbol{v} = v\,\hat{\boldsymbol{e}}_t$ and $\boldsymbol{a} = v^2/r\,\hat{\boldsymbol{e}}_n$. Here, v is the orbital speed, $\hat{\boldsymbol{e}}_t$ is tangent to the orbital path, r is the distance separating the bodies, and $\hat{\boldsymbol{e}}_n$ is along the line joining the two bodies. The distance r is commonly expressed as $r = R + h$ where R is the mean radius of the large spherical body and h is the orbit altitude.

The acceleration of the orbiting body is due to the gravitational attraction, however, and Newton's laws of motion combine to give that the amount of gravity at an altitude h above the large body's surface is $g_0 R^2/(R + h)^2$. Consequently, $\boldsymbol{a} = g_0 R^2/(R + h)^2 \hat{\boldsymbol{e}}_n$. Here, g_0 is the gravitational acceleration constant at the surface of the large body.

One finds, then, that the constant speed v can be related to the orbit altitude.

$$v = R\sqrt{g_0/(R + h)} \tag{39.1}$$

As an example calculation, for a satellite in *geostationary orbit* ($h = 22,240$ mi) around Earth ($g_0 = 32.2$ ft/sec^2; $R = 3963$ mi), the satellite speed is nearly 2 mi/sec.

Cylindrical Coordinates for Motion in 3-D. Position, velocity, and acceleration vectors in 3-dimensions can be expressed using cylindrical coordinates along a rotating frame. Consider a reference frame that rotates like before: $\hat{e}_1$ and $\hat{e}_2$ rotate in a horizontal plane whereas $\hat{e}_3$ always points vertical. The coordinates (r, θ, z) are polar-Cartesian mix.

$$\boldsymbol{p} = r\hat{e}_1 + z\hat{e}_3 \tag{40.1}$$

$$\boldsymbol{v} = \dot{r}\hat{e}_1 + r\dot{\theta}\hat{e}_2 + \dot{z}\hat{e}_3 \tag{40.2}$$

$$\boldsymbol{a} = (\ddot{r} - r\dot{\theta}^2)\hat{e}_1 + (r\ddot{\theta} + 2\dot{r}\dot{\theta})\hat{e}_2 + \ddot{z}\hat{e}_3 \tag{40.3}$$

★ 8 The Spiraling Descent of a Glider.

A glider descends in a perfectly helical path such that the altitude $h = r/2\,\theta$ where $r = 2m$. The descent rate is a constant 2m per 5 sec. Determine the kinematics using cylindrical coordinates. Compute $\underline{v}$ and $\underline{a}$.

The gliding particle

Begin with the position vector.

$$\underline{r} = r\,\hat{e}_1 + h\,\hat{e}_3$$
$$= r\,\hat{e}_1 + r/2\,\theta\,\hat{e}_3$$

Next determine the velocity vector.

$$\underline{v} = r\dot{\theta}\,\hat{e}_2 + \dot{h}\,\hat{e}_3 = r\dot{\theta}\,\hat{e}_2 + r/2\,\dot{\theta}\,\hat{e}_3$$

Next determine the acceleration vector.

$$\underline{a} = -r\dot{\theta}^2\,\hat{e}_1 + r\ddot{\theta}\,\hat{e}_2 + r/2\,\ddot{\theta}\,\hat{e}_3$$

Perform some computations

$$\dot{\theta} = 2/r\,\dot{h} = 2/2 \cdot 2/5 = 2/5\;\frac{rad}{sec}$$
$$\ddot{\theta} = 0$$
$$\underline{v} = (4/5\,\hat{e}_2 + 2/5\,\hat{e}_3)\;m/sec$$
$$\underline{a} = -8/25\,\hat{e}_1\;m/sec^2$$

Chapter 3 Problem Set.

1. (M&K p.72) A tracking radar station is able to measure range and tilt away from the local vertical direction, as shown in the illustration. Suppose the radar tracks a gliding decoy that moves in a vertical plane. Because the decoy is unpowered, its acceleration is due to gravity only and is given by $a = 31.4$ ft/sec^2 at the current altitude. The measurements at a particular instant are $\theta = 30$ deg; $r = 25(10^4)$ ft; $\dot{r} = 3000$ ft/sec; and $\dot{\theta} = 0.80$ deg/sec. (a) Establish a stationary reference frame and a rotating reference frame that can be used to describe the motion of the decoy. (b) Write kinematic equations for the acceleration, velocity, and position vectors using Cartesian coordinates with components along the stationary frame directions. (c) Write kinematic equations for the acceleration, velocity, and position vectors using polar coordinates with components along the rotating frame directions. (d) Compute the instantaneous speed of the decoy. (e) Show that $\ddot{r} = r\dot{\theta}^2 + a\cos\theta$ and compute the instantaneous value of $\ddot{r}$. (f) Compute the instantaneous value of $\ddot{\theta}$. Answer: $v = 4602$ ft/sec.

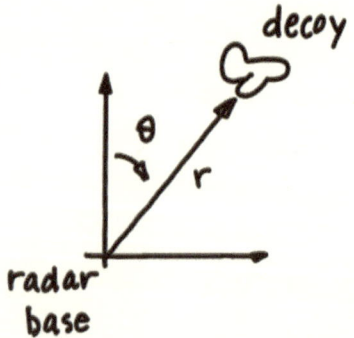

2. (M&K 2.147) A rocket is fired in the vertical direction. It experiences a velocity and acceleration along a straight vertical path. A radar station that measures range and elevation tracks the liftoff and subsequent ascent. Data is recorded at the instant $\theta = 60$ deg: $r = 32{,}000$ ft; $\ddot{r} = 70$ ft/s^2; and $\dot{\theta} = 0.03$ rad/s. (a) Write kinematic equations for the acceleration, velocity, and position vectors using polar coordinates with components along rotating frame directions. (b) Write kinematic equations for the acceleration, velocity, and position vectors using Cartesian coordinates with components along stationary frame directions. (c) Compute the magnitudes of the velocity and acceleration vectors at this instant.

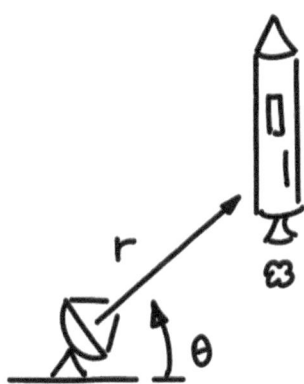

3. (M&K 2.167) A meteor chunk is tracked by a radar station as it flies overhead. The radar station is able to measure range r, range rate $\dot{r}$, elevation θ, and elevation rate $\dot{\theta}$. Assume a non-rotating Earth in the following developments. (a) Write an expression for the velocity vector using polar coordinates with components along the stationary $\hat{n}_1$ and $\hat{n}_2$ reference directions. (b) Show that the angle β, which the velocity vector makes with the horizontal direction, is given by the solution to $\tan \beta = (\dot{r} \sin \theta + r\dot{\theta} \cos \theta)/(\dot{r} \cos \theta - r\dot{\theta} \cos \theta)$. (c) Suppose the radar measurements are $\theta = 90$ deg; $r = 90$ km; $\dot{r} = -20$ km/s; and $\dot{\theta} = 0.4$ rad/s. Compute the speed of the debris and compute β. (d) Repeat the computations if $\theta = 75$ deg. Answers: $v = 41.1$ km/sec ; $\beta = 29$ deg.

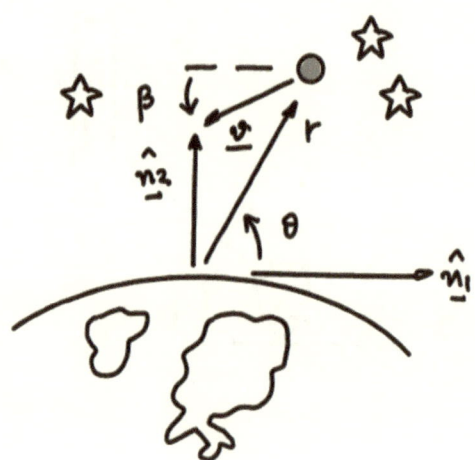

4. (M&K 2.168) A bottle rocket is launched vertically into the air. Because of drag, the rocket experiences a vertical acceleration of $a = -g - kv^2$, where k is a known drag coefficient and v is the known instantaneous vertical speed. (a) Find expressions that relate r, $\dot{r}$, $\ddot{r}$, θ, $\dot{\theta}$, and $\ddot{\theta}$ to the known instantaneous vertical speed and other constants. (b) Evaluate your expressions for $k = 0.012/m$ and $v = 16$ m/s. Answer: $\ddot{r} = -4.8$ m/sec^2.

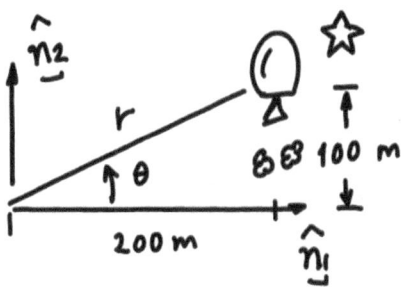

5. A rapid response satellite is launched from a rail track, which is shown in the figure as the dashed line. The track makes a constant angle ϕ relative to the $\hat{n}_1$ direction. This angle is known. A measurement system is located at the origin o, and the distance b from o to o' is known. The measurement device can output θ and its time derivatives. (a) Compute the speed of the satellite as it travels along the track in terms of b, ϕ, θ, and the time derivatives of θ. (b) Compute the change in speed of the satellite as it travels along the track in terms of b, ϕ, θ, and the time derivatives of θ.

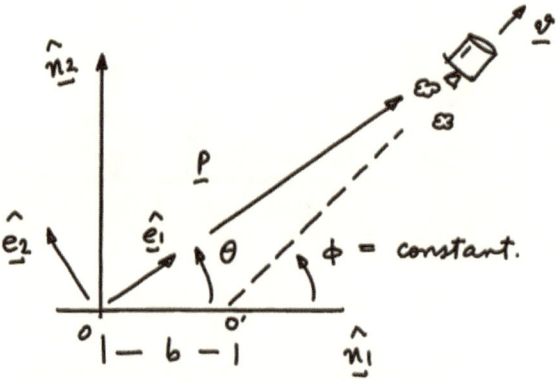

6. A projectile will touch its trajectory envelope sometime during its flight if the launch angle is between 45 and 90 degrees. The time to hit the envelope is $t_e = v/(g \sin \gamma)$ where v is the known launch speed and γ is the launch angle as measured from the horizontal. (a) Show that an expression for the range rate, as measured from the launch point, at the instant when the projectile touches its trajectory envelope is $\dot{r} = v \cos^3 \gamma \, (3 - 4 \cos^2 \gamma)/ \sin \gamma$. (b) What value of γ maximizes this expression for $\dot{r}$? (c) Plot $\dot{r}$ as a function of γ for a set value of v.

7. (M&K 2.159) A space shuttle travels in a circular orbit around the Earth at an altitude of h miles. A radar station located at o accurately measures the range rate and elevation angle at a particular instant. The radius of the Earth is approximately $R = 3959$ miles. (a) Show that the magnitude of the orbital speed of the shuttle is given by the expression $v = \dot{r}(R + h)/(R \cos \theta)$.

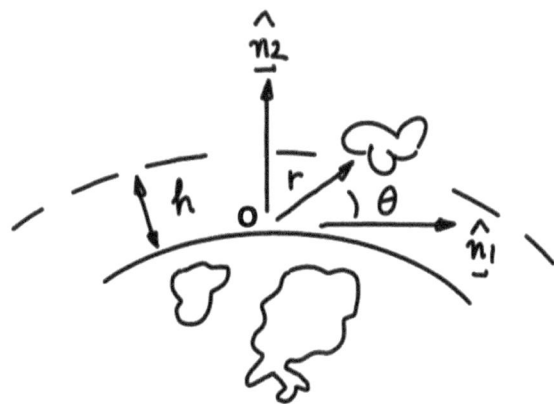

Chapter 3 Notes.

4 Point Mass Kinetics in Stationary Frames

An Introduction. Find. Analyze. Solve.[6] Find the equations of motion, analyze them to glean important traits, and solve them to know detailed information.

These steps partially describe the job of a dynamicist. The zeroth step is to mathematically model the actual system, but we won't do much modeling here. We won't do much analysis either. We'll primarily focus on finding the equations of motion and solving them to some degree.

Newton gave us $f = m\,a$, and this is what we use to find the equations of motion. We'll develop and practice this skill here.

---

[6] From the well-known mnemonic $\int f = \int ma \rightarrow$ *f*ind, *a*nalyze, *f*olve.

Newton's Second Law: N2L. Newton's three laws of motion for a point mass lead to the important result $\boldsymbol{f} = m\,\boldsymbol{a}$.

The left side of $\boldsymbol{f} = m\boldsymbol{a}$ contains the vector sum of the impressed forces on the point mass. Identifying the vector sum $\boldsymbol{f}$ is facilitated by the use of a free body diagram (FBD). When drawing a FBD, it is important to include only those forces that are caused by contact (e.g., the point mass is touching a wall), or impressed by some external device (e.g., a spring is attached to the point mass), or generated by a vector field (e.g., gravity). "Inertia forces" or "centrifugal forces" or "centrifical forces" are not impressed forces and are not included: these fictitious forces are properly accounted for in the kinematics.

The right side of $\boldsymbol{f} = m\,\boldsymbol{a}$ is all about the inertial acceleration vector of the point mass. Inertial acceleration of a point mass is the acceleration vector witnessed by an inertial observer. The inertial acceleration vector is the time derivative of the inertial velocity vector $\boldsymbol{v}$, which is the time derivative of the inertial position vector $\boldsymbol{p}$. Obtaining the correct governing equations of motion (which is what $\boldsymbol{f} = m\,\boldsymbol{a}$ produces) directly hinges on performing the correct vector kinematics. We will discover that wise selections of references frames and generalized coordinates are often beneficial.

A Point Mass Routine. Newton's second law generates second-order differential equations that govern the motion of a point mass. A procedure for generating the equations can be established.

1. Identify the system.

2. Draw a free body diagram (FBD).

3. Establish relevant reference frames.

4. Write a vector representation of the forces.

5. Perform the vector kinematics up to the acceleration level.

6. Use N2L to write the governing equations of motion.

⋆ 9 A Block on an Inclined Plane.

Consider a block on an inclined plane. The block is acted on by a force T. There is no friction between the contact surfaces. Gravity acts downward. The mass, force T, and angle θ are known. Compute the acceleration $\underline{a}$.

Steps 1 & 2.
The system
& FBD.

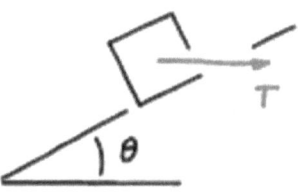

Step 3.
Relevant reference frame.

Step 4.
Vector representation of the forces

$$\underline{f} = \underline{T} + \underline{w} + \underline{N} = T\cos\theta \; \hat{b_1} - T\sin\theta \; \hat{b_2}$$
$$- W\cos\theta \; \hat{b_2} - W\sin\theta \; \hat{b_1} + N \; \hat{b_2}$$

step 5. Kinematics

$$\underline{P} = x \; \hat{b_1}, \quad \underline{V} = v \; \hat{b_1}$$
$$\underline{a} = a \; \hat{b_1}$$

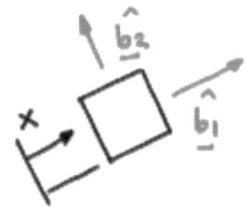

step 6. N2L $\underline{f} = m\underline{a}$. Equate $\hat{b_1}$ & $\hat{b_2}$ parts

$$T\cos\theta - mg\sin\theta = ma$$
$$-T\sin\theta - mg\cos\theta + N = 0$$

Easily solve for a & N.

⋆ 10 A Block on an Inclined Plane II.

The previous problem used reference vectors that were aligned with the incline. Suppose this was not the case.

Steps 1 & 2 don't change
step 3. Reference frame

Step 4. Vector representation of forces

$$\underline{f} = T \,\hat{n_1} - mg \,\hat{n_2} - N \sin\theta \,\hat{n_1} + N \cos\theta \,\hat{n_2}$$

step 5. kinematics

$$\underline{a} = a \cos\theta \,\hat{n_1} + a \sin\theta \,\hat{n_2}$$

Step 6. N2L. $\underline{f} = m\underline{a}$. Equate $\hat{n_1}$ & $\hat{n_2}$ parts.

$$T - N \sin\theta = m a \cos\theta$$
$$N \cos\theta - mg = m a \sin\theta$$

The acceleration a & normal N are coupled in this version; they are not isolated in separate equations like before. These equations give the same answers, of course.

⋆ 11 A Collar on a Shaft.

consider a collar that can slide on a vertical shaft. The collar is acted on by a force T. There is no friction between the contact surfaces. Gravity acts downward. The mass, force T and θ are known. Determine the acceleration.

Step 1.

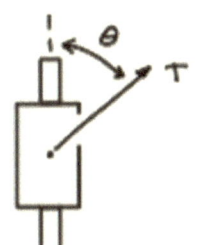

step 2.

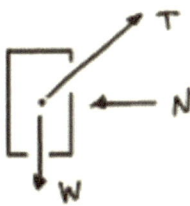

Step 3.

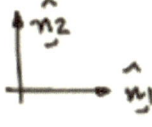

step 4.

$$\underline{f} = \underline{T} + \underline{W} + \underline{N}$$
$$= T\cos\theta\ \hat{n}_2 + T\sin\theta\ \hat{n}_1$$
$$- W\ \hat{n}_2 - N\ \hat{n}_1$$

Step 5.

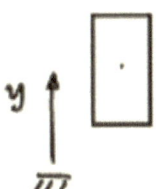

$$\underline{p} = y\ \hat{n}_2 \qquad \underline{v} = v\ \hat{n}_2$$

$$\underline{a} = a\ \hat{n}_2$$

Step 6. N2L: $\underline{f} = m\underline{a}$

$$T\cos\theta - mg = ma$$
$$T\sin\theta - N = 0$$

⎫
⎬ Easily find a & N
⎭

Equations of Motion: Now What? Newton's second law generates second-order differential equations that govern the motion of a point mass. A procedure for generating the equations was established. Point mass motion in one variable leads to one equation for the acceleration.

$$ma = f(t, x, v) \qquad \text{or} \qquad m\dot{v} = f(t, x, v) \tag{56.1}$$

At this point one could be given the acceleration and asked to compute the instantaneous force that is consistent with the acceleration. Or one could be given the force and asked to compute the instantaneous acceleration. Either way, one is left with an algebra problem.

If f depends on only one of the variables t, x, or v, then integral calculus can be used to dig deeper. Some examples are presented in the following pages for illustration.

1. $f = f(t)$ is known and we find velocity and position as a function of time.

2. $f = f(x)$ is known and we find velocity as a function of position. Here $f(x)$ is a linear function of position.

3. $f = f(x)$ is known and we find velocity as a function of position. Here $f(x)$ is a nonlinear function of position.

4. $f = f(v)$ is known and we find position as a function of velocity. Here $f(v)$ is a linear function of velocity.

5. $f = f(v)$ is known and we find time as a function of velocity, and then invert the relationship to find velocity in terms of time. Here $f(v)$ is a nonlinear function of velocity.

⋆ 12 A Model Rocket Problem.

A 100 gram model rocket is launched vertically from rest. A constant thrust 10 N acts for 1 second. There is no thrust afterward. Gravity acts downward. Compute the maximum vertical height.

steps 1, 2, 3, 4.

$$\underline{F} = -mg \ \hat{n_2} + T \ \hat{n_2} \qquad \text{mass}$$

{ The thrust only acts for 1 second. }

step 5.

$$\underline{p} = \gamma \ \hat{n_2} \qquad \underline{v} = v \ \hat{n_2} \qquad \underline{a} = a \ \hat{n_2}$$

step 6. $\underline{f} = m\underline{a}$

$$-mg + T = ma = ma = m \ dv/dt = m \ \frac{d^2y}{dt^2}$$

Solving for general expressions

$$v = (T/m - g) \ t \qquad y = \frac{1}{2} (T/m - g) \ t^2$$

At time t = 1 sec.

$$v_1 = (T/m - g) \qquad y_1 = \frac{1}{2} (T/m - g)$$

For time > 1 second, N2L gives $-mg = m \ dv/dt$
The velocity expression for t > 1 is

$v(t) = v_1 - g(t-1)$. Another relationship is

$v^2 - v_1^2 = -2g(y - y_1)$. The max height is

$$y_{max} = v_1^2/2g + y_1 = T/2m (T/mg - 1)$$

★ 13 The Archer's Bow.

The string from an archer's bow exerts a force on an arrow that nearly obey's Hooke's law, $f = \kappa x$. Find the governing equation of motion & compute the velocity as the arrow leaves the bow string.

Steps 1, 2, 3, 4.

Step 5.

$$\underline{P} = x \, \hat{n}_1$$

$$\underline{v} = v \, \hat{n}_1 \qquad \underline{a} = a \, \hat{n}_1$$

Step 6. $\underline{f} = m\underline{a}$

$$-\kappa x = ma = m \, dv/dt$$

Using the chain rule

$$dv/dt = dv/dx \; dx/dt = dv/dx \; v$$

Thus $\quad -\kappa/m \; x \; dx = v \, dv$

Solving $-\kappa/m \left(\dfrac{x^2}{2} - \dfrac{x_0^2}{2} \right) = \dfrac{v^2}{2} - \dfrac{v_0^2}{2}$

Evaluating with $v_0 = 0$ and $x = 0$,

$$\dfrac{k}{m} x_0^2 = v^2 \quad \text{or} \quad v = x_0 \sqrt{\kappa/m}$$

For example, if $k = 200 \; \dfrac{N}{m}$ & $m = 20$ grams & $x_0 = 3/4$ m

Then $v \approx 170 \; \dfrac{\text{miles}}{\text{hr}}$

⋆ 14 A Point Mass Lunar Rocket.

A rocket is launched directly away from a heavenly body using constant thrust. Use Newton's inverse square law to model gravity. Assume the rocket begins from rest at the surface. Find the governing equation of motion ; compute the velocity as a function of altitude.

Steps 1, 2, 3, 4.

$$\underline{F} = (T - G) \, \hat{\underline{n}}_1 = T - mg_0 \frac{R^2}{r^2}$$

Step 5. $\quad \underline{r} = r \, \hat{\underline{n}}_1 \qquad \underline{v} = v \, \hat{\underline{n}}_1 \qquad \underline{a} = a \, \hat{\underline{n}}$

Step 6. $\quad \underline{f} = m \underline{a}$

$$T - mg_0 \frac{R^2}{r^2} = ma = m \, dv/dt$$

Using the chain rule

$$dv/dt = dv/dr \; dr/dt = dv/dr \; v$$

$$T/m \, dr - g_0 R^2/r^2 \, dr = v \, dv$$

$$T/m \, (r - R) + g_0 R^2 \left(1/r - 1/R \right) = v^2/2$$

$$v = \sqrt{ \frac{2T}{m} (r - R) - g_0 \frac{R}{r} (r - R) }$$

⋆ 15 A Point Mass in a Thick Fluid.

A point mass is released from rest into a thick fluid. The fluid resists the motion by a force that is proportional to the speed. Find the governing equation of motion & compute the depth as a function of velocity.

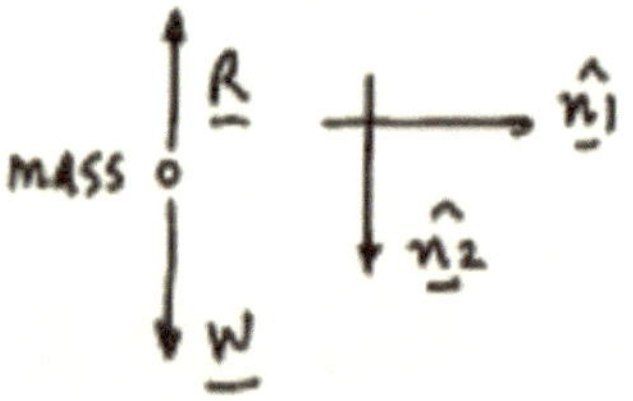

steps 1, 2, 3, 4.

$$\underline{F} = mg \, \hat{n_2} - R \, \hat{n_2}$$

$$= mg \, \hat{n_2} - bv \, \hat{n_2}$$

steps 5.

$$\underline{r} = y \, \hat{n_2} \qquad \underline{v} = v \, \hat{n_2} \qquad \underline{a} = a \, \hat{n_2}$$

steps 6. $\underline{f} = m\underline{a}$; $\qquad mg - bv = ma$

Using the chain rule

$$a = dv/dt = \frac{dv}{dy} \frac{dy}{dt} = \frac{dv}{dy} v$$

So

$$mg - bv = m \frac{dv}{dy} v \qquad or \qquad dy = \frac{v \, dv}{g - bv/m}$$

Integrating

$$y = \frac{g}{c^2} \ln \left(\frac{1}{1 - cv/g} \right) - v/c \qquad where \quad c \equiv \frac{b}{m}$$

{ The fluid drag law holds for small v. }

⋆ **16 The Sky Diver.**

A sky diver experiences a gravity force and an aero-dynamic drag force. At high velocities the drag force is proportional to the speed squared. Find the governing equation of motion & compute the speed as a function of time.

Steps 1, 2, 3, 4.

$$\underline{F} = -mg \ \hat{n2} + D \ \hat{n2}$$

$$= -mg \ \hat{n2} + bv^2 \ \hat{n2}$$

mass

Step 5. $\underline{P} = y \ \hat{n2}$; $\underline{v} = v \ \hat{n2}$ $\underline{a} = a \ \hat{n2}$

Step 6. $\underline{F} = m\underline{a}$; $-mg + bv^2 = ma = m \ dv/dt$

A diver will experience a terminal velocity v_* when $a = 0$, which implies $b \, v_*^2 = mg$. Thus, the equation of motion can be re-arranged.

$$-\frac{g}{v_*^2} \, dt = \frac{dv}{(v_*^2 - v^2)}$$

Integrating

$$\frac{1}{2v_*} \ell n \left(\frac{v_* + v}{v_* - v} \right) = \frac{-g}{v_*^2} \, t$$

or $\quad v = -v_* \dfrac{1 - \exp(-2gt/v_*)}{1 + \exp(-2gt/v_*)}$

{ what's the solution for x ? }

Euler's Simple, Simple Numerical Integration.   Newton's second law for a point mass motion in one variable leads to one equation for the acceleration.

$$ma = f(t, x, v) \qquad \text{or} \qquad m\dot{v} = f(t, x, v) \tag{62.1}$$

If f depends on more than one of the variables t, x, or v, then analytical solutions may be difficult to find. One alternative is to use a simple numerical technique. Consider the following idea.

At the initial time, the acceleration can be found in terms of the initial conditions, $a_0 = f(t_0, x_0, v_0)/m$. Consider a short time interval Δt, or more simply h, and suppose a_0 is considered constant over this time interval. Then the next velocity can be computed.

$$\frac{v_1 - v_0}{h} = a_0 \qquad \text{or} \qquad v_1 = v_0 + a_0 h \tag{62.2}$$

In a similar way, suppose the next position can be computed based on the now known velocities.

$$\frac{x_1 - x_0}{h} = v_{\text{avg}} = \frac{v_1 + v_0}{2} \qquad \text{or} \qquad x_1 = x_0 + v_0 h + a_0 h^2/2 \tag{62.3}$$

The next time is simply $t_1 = t_0 + h$.

Now with t_1, x_1, and v_1 known, the next acceleration can be computed, $a_1 = f(t_1, x_1, v_1)/m$. And round and round we go.

$$a_{n-1} = f(t_{n-1}, x_{n-1}, v_{n-1})/m \tag{62.4}$$

$$v_n = v_{n-1} + a_{n-1} h \tag{62.5}$$

$$x_n = x_{n-1} + v_{n-1} h + a_{n-1} h^2/2 \tag{62.6}$$

⋆ 17 An Euler Numerical Integration Picture.

An example of the Euler numerical integration routine is shown here. The solid line is the approximate numerical integration and the dashed line is the exact solution. This simple approach to numerically solve for the motion works well some times.

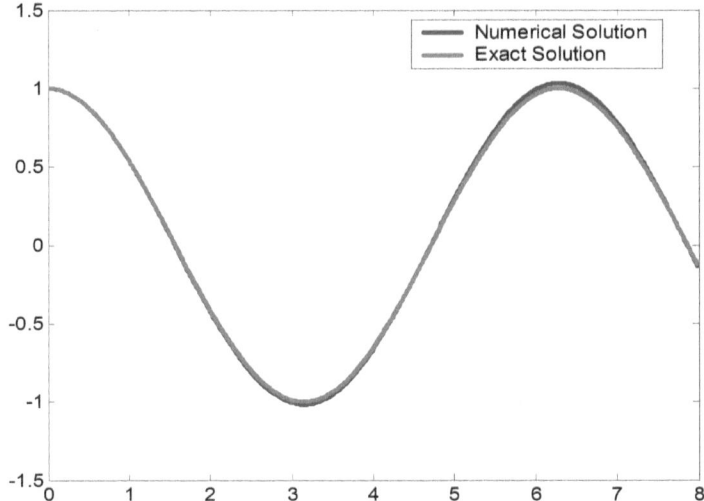

Rigidly Connected Bodies. When two or more point masses are rigidly connected together, then Newton's second law can be applied to each mass or it can be applied to the overall system. Or one can do both.

Significantly, when one considers the system as a whole, then the connection force between the point masses is not exposed. This can be useful if it is not known and its value is not important. If the connection force between the point masses is known or needs to be known, then $f = ma$ can be applied to each point mass to help determine its value.

Importantly, a rigid connection between bodies means that their accelerations will be related.

These ideas are shown in the following pages where we first consider 2 bodies, then n bodies, then an ∞ number of bodies.

★ 18 Two Blocks.

Blocks A & B are connected
by a cable that remains taut.
Find the equations of motion.
Compute the acceleration of A
and the tension in the cable.

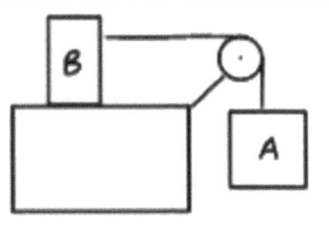

steps 1, 2, 3, 4.

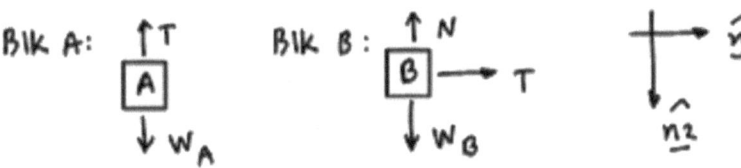

Forces on A: $\underline{f}_A = (-T + W_A)\,\hat{\underline{n}}2$

Forces on B: $\underline{f}_B = (-N + W_B)\,\hat{\underline{n}}2 + T\,\hat{\underline{n}}1$

Step 5.

Kinematics of A:

$\underline{r}_A = X_A\,\hat{\underline{n}}2$

$\underline{V}_A = \underline{\vartheta}_A\,\hat{\underline{n}}2$

$\underline{a}_A = a_A\,\hat{\underline{n}}2$

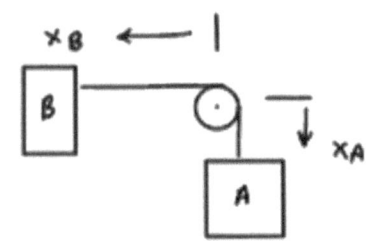

Kinematics of B:

$\underline{r}_B = -X_B\,\hat{\underline{n}}1$

$\underline{V}_B = -\underline{\vartheta}_B\,\hat{\underline{n}}1$

$\underline{a}_B = -a_B\,\hat{\underline{n}}1$

NOTE: There is a kine-
matic constraint.

$X_B + X_A = L$

so, $a_B = -a_A \equiv -a$

step b.

N2L A : $-T + M_A g = m_A a_A = m_A a$

N2L B : $m_B g = N$

$T = -m_B a_B = m_B a$

Solving $m_A g = (M_A + M_B) a$

$$a = \frac{M_A g}{M_{total}}$$

$$T = \frac{m_B M_A}{M_{total}} g$$

Example $M_A = 25 \, kg$ $m_B = 30 \, kg$

$g = 9.81 \, m/s^2$

$a = 4.46 \, m/s^2$; $T = 133.8 \, N$

★ 19 A Toy Train.

A toy train is held together by magnetic couplers. Each coupler can provide a maximum force of T_*. Which coupler will fail first? What is the maximum pull force $\underline{T}$?

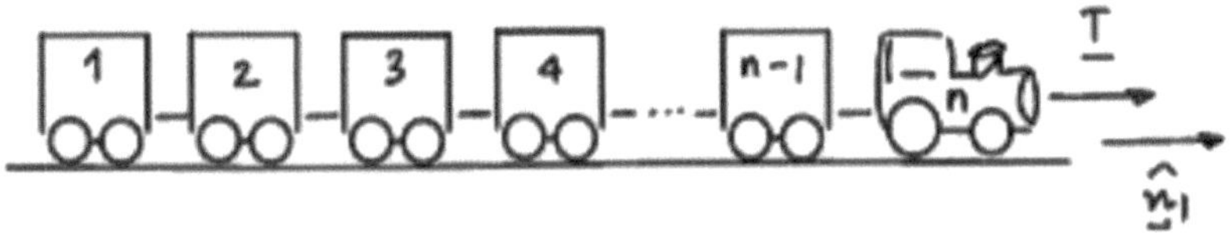

Investigate car #1:

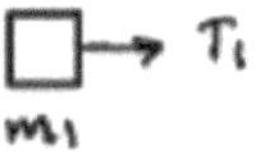

$\square \rightarrow T_1$
m_1

$T_1 = m_1 a_1$

Investigate car #2:

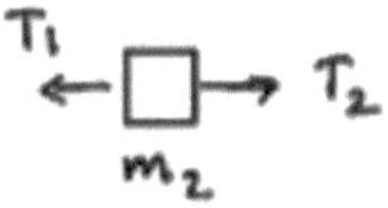

$T_1 \leftarrow \square \rightarrow T_2$
m_2

$T_2 - T_1 = m_2 a_2$

Assume the train cars accelerate together so $a_1 = a_2$. Then

$$T_2 = T_1 + m_2 a = (m_1 + m_2) a$$
$$= \sum_{j=1}^{2} m_j \, a$$

Investigate car #n-1:

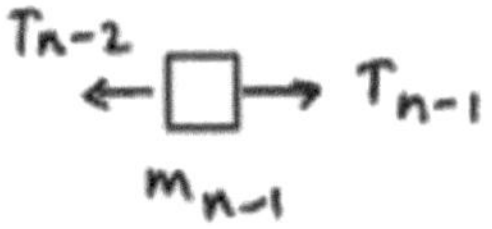

$T_{n-2} \leftarrow \square \rightarrow T_{n-1}$
m_{n-1}

$T_{n-1} - T_{n-2} = m_{n-1} a$

One can realize

$$T_{n-1} = \sum_{j=1}^{n-1} m_j \, a$$

So $T_{n-1} > T_{n-2} > \cdots T_2 > T_1$

So T_{n-1} is largest and it will break first.

Let $T_{n-1} = T_*$.

So $T_* = \sum_{j=1}^{n-1} m_j \, a = (m_{total} - m_n) \, a$

so $a = \dfrac{T_*}{(m_{total} - m_n)}$ This is the largest permissible acceleration

NOW compute permissible force T by investigating entire train.

$m_{total} \rightarrow T$ $T = m_{total} \, a$

so $T = \dfrac{T_* \, m_{total}}{(m_{total} - m_n)}$

⋆ 20 A Hanging Chain.

A chain with length L is released from rest with an amount C hanging over the edge of a horizontal table. Assume the chain slides without friction. Compute the velocity as the end leaves the table.

Chain density

$\rho = \dfrac{mass}{length}$

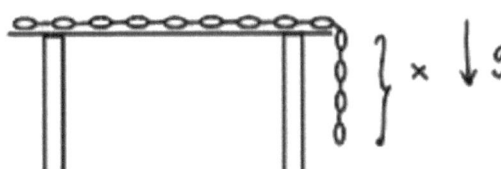

$\Big\} \times \downarrow g$

Let's quickly investigate a hanging portion using the six steps.

$\uparrow T$

$\downarrow W \quad \downarrow \hat{n_2}$

$\rho \times g - T = \rho \times a$

Let's quickly investigate the portion remaining on the table.

$\uparrow N$

$\leftarrow T \qquad \rightarrow \hat{n_j}$

$\downarrow W \qquad \downarrow \hat{n_2}$

$N = \rho(L-x)g$

$T = \rho(L-x)a$

These equations reveal $\rho g x = \rho L a$

Integrating using $a = dv/dt = v\, dv/dx$

$$\int_{c}^{L} \frac{g}{L} x\, dx = \int_{0}^{v} v\, dv \quad ; \quad \frac{g}{L}(L^2 - c^2) = v^2$$

$$v = \sqrt{\frac{g}{L}(L-c)(L+c)} \quad ; \quad v \to \sqrt{gL} \quad \text{as} \quad c \to 0$$

Integrated Or Momentum Form.   Some texts mention a momentum form, or integrated form, or finite time or displacement form of Newton's second law.

What they have in mind is $\int \boldsymbol{f} = m \int \boldsymbol{a}$ instead of $\boldsymbol{f} = m\boldsymbol{a}$ (70.1)

If the left expression is an integrated form, then I suppose the right expression is a differential form. We've already done some work with the integrated form, but we'll revisit it here.

Newton's second law for a point mass is the vector equation $\boldsymbol{f} = m\boldsymbol{a}$. We've been using a standard approach for finding the governing equations of motion, which involves identifying the system, drawing FBDs and introducing reference frames, writing a vector representation of the forces, writing the vector kinematics up to the acceleration level, and finally using N2L. We've seen that solving the equations, however, depends on the form of $\boldsymbol{f}$. We've found some analytical expressions when $\boldsymbol{f}$ depends on a single variable, e.g., $\boldsymbol{f}(t)$ or $f(x)$.

When $\boldsymbol{f} = \boldsymbol{f}(t)$, we have a momentum equation.

$$\int_{t_0}^{t_1} \boldsymbol{f}(t) \, \mathrm{d}t = m \int_{v_0}^{v_1} \mathrm{d}\boldsymbol{v} = m(\boldsymbol{v}_1 - \boldsymbol{v}_0) \qquad (70.2)$$

The quantity $m\boldsymbol{v}$ is the translational momentum of the point mass. So the change in the point mass momentum over the interval $\Delta t = t_1 - t_0$ can be computed if $\boldsymbol{f}(t)$ can be integrated.

When $f = f(x)$, we have an energy equation.

$$\int_{x_0}^{x_1} f(x) \, \mathrm{d}x = m \int_{v_0}^{v_1} v \, \mathrm{d}v = \frac{m}{2}(v_1^2 - v_0^2) \qquad (70.3)$$

The quantity $(m/2)v^2$ is the kinetic energy of the point mass. So the change in the point mass kinetic energy over the interval $\Delta x = x_1 - x_0$ can be computed if $f(x)$ can be integrated.

As stated, we've already done this kind of thing (see the Model Rocket Problem and the Archer's Bow Problem), but formally seeing things in this way allows us to discover the concepts of translational momentum and kinetic energy.

⋆ 21 Cart Momentum.

A cart is subject to a horizontal time-varying force P. The cart begins from rest. Compute the translational momentum at $t = 6$ seconds.

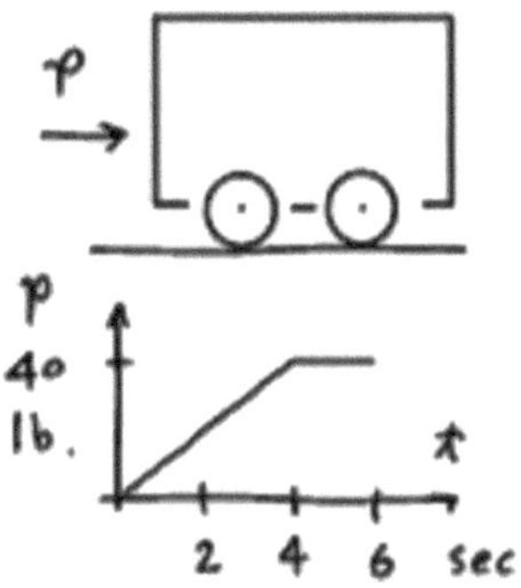

Steps 1,2,3,4.

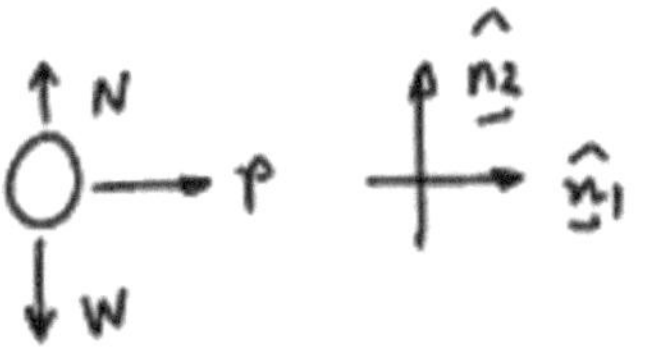

$$\underline{F} = (N - W)\,\hat{n}_2 + P\,\hat{n}_1$$

step 5. $\underline{r} = x\,\hat{n}_1$; $\underline{v} = v\,\hat{n}_1$; $\underline{a} = a\,\hat{n}_1$

step 6. $\underline{F} = m\underline{a}$: $N = W$; $P = ma$

Solving

$$\int_0^6 P(t)\,dt = m\,v - m\,v_0 = m\,v$$

$$\int_0^4 10\,t\,dt + \int_4^6 40\,dt = m\,v$$

$$10 \cdot \frac{16}{2} + 40 \cdot 2 = m\,v$$

so $m\,v = 160$ lb·sec

Chapter 4 Problem Set.

1. A 2 kg point mass is suspended from the ceiling by a linear spring. The spring constant is $k = 20$ N/m, and for simplicity, let the unstretched natural length of the spring be zero. Gravity acts downward and let $g = 10$ m/s². (a) Follow the usual 6 steps to develop the governing equations of motion. (b) Suppose the mass begins from rest at $x = 2$ meters. Compute the speed when $x = 1$ meter.

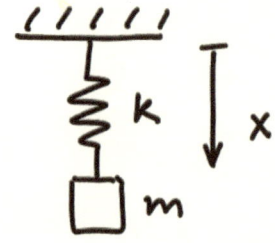

2. Use MATLAB or something similar to numerically simulate the vertical motion of a sky diver. The equations of motion were presented earlier in this chapter. Use Euler's Simple Numerical Integration technique. Use the following parameters: final time = 30 sec; initial altitude 1400 m ; mass = 70 kg ; terminal velocity = 54 m/s. (a) Plot the altitude versus time resulting from the numerical simulation together with the exact solution. (b) Plot the vertical velocity versus time resulting from the numerical simulation together with the exact solution.

3. A rocket test apparatus is modeled as shown in the right figure. A linear spring with constant k attaches the mass m to a stationary wall. The spring is a mathematical spring, and it always only acts horizontally. A constant thrust T acts horizontally to the right. A damper with constant b opposes vertical velocity, which means that if the vertical speed is up, the damper force is down with magnitude bv_y. The system of concern is the block m, which can move in horizontal and vertical directions in the plane. Assume that gravity can be neglected. (a) Follow the usual 6 steps to develop the governing equations of motion. (The one vector equation $f = ma$ produces two scalar equations for motion in the plane.) (b) Solve the scalar vertical equation for the instantaneous vertical speed v_y as a function of time given that the mass has an upwards speed v_{y0} at time $t = 0$. (c) Solve the scalar horizontal equation for an expression that relates the instantaneous horizontal speed v_x to the instantaneous horizontal position x and thrust T given that the initial horizontal position and horizontal speed are both zero at time $t = 0$.

74

4. Long, skinny rockets in flight can experience longitudinal elastic vibra-
tions. Consider a simple mathematical model of a long, skinny rocket,
consisting of two masses connected to each other with a linear spring.
The rocket experiences flight along a vertical direction. The thrust is
constant. The un-stretched length of the linear spring is ℓ. (a) Use the
six-step approach to derive the equations of motion for mass A. (b) Use
the six-step approach to derive the equations of motion for mass B. (c)
Define a relative coordinate between the two masses, $z = x_A - x_B$. There-
fore $\dot{z} = \dot{x}_A - \dot{x}_B$ and $\ddot{z} = \ddot{x}_A - \ddot{x}_B$. Using the previous results, find
an equation of motion in terms of the relative coordinate z. This is an
equation that governs the relative motion. (d) Use the chain rule to solve
the relative motion equation, which will relate the current relative position and velocity to the initial relative position and velocity. (e) Find an
expression for the center of mass. Find the governing equation of motion
for the center of mass. Completely solve the center of mass equation of
motion.

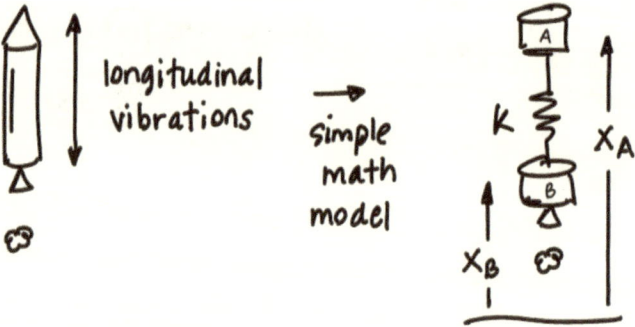

5. A model test of the Advanced Hypersonic Weapon (AHW) is conducted at Sandia Laboratories. The rocket is treated as a point mass of constant mass m. The rocket is subject to a constant vertical thrust T_1 Newtons until time t_1 seconds and then a constant vertical thrust T_2 Newtons until time t_2 seconds. All motion takes place along the vertical direction and gravity acts downward. The rocket begins from rest on the launch pad. (a) Derive an expression that gives the rocket speed at the end of the second burn in terms of the thrusts, weight, times, etc. (b) What is the sensitivity (or derivative) of the speed at the end of the second burn with respect to the thrust T_2?

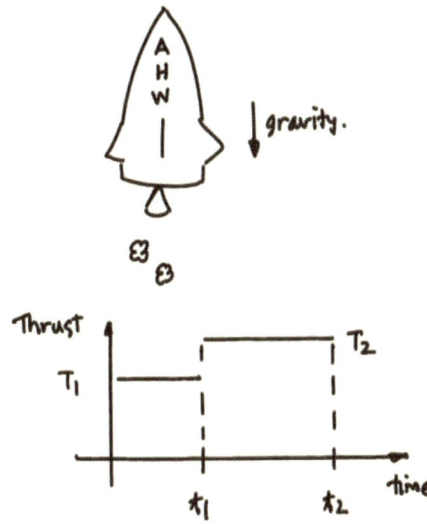

6. This is a classic problem. Consider the two mass system shown. We seek the smallest deflection of the top mass, which when released from a rest position causes the bottom mass to almost achieve lift-off. Assume that all motion occurs in the vertical direction. Gravity acts downward, and the spring has an unstretched length of ℓ. Note that the magnitude of the spring force is $T = k(x_1 - x_2 - \ell)$, where x_1 is the inertial position of the top mass and x_2 is the inertial position of the bottom mass. Our mathematical model will be valid so long as x_2 remains unchanged, which means $\dot{x}_2 = \ddot{x}_2 = 0$ for all time. (a) Use the six-step approach to derive the equations of motion for the top mass. In doing so, define $x = x_1 - x_2 - \ell$ and express your differential equation of motion in terms of the variable x. (b) Use the chain rule to integrate the equation of motion of the top mass. When evaluating the expression at the limits of integration, take advantage of the fact that the top mass begins from rest at x_0. Furthermore, evaluate the expression when the top mass achieves zero velocity, and denote this position as x_*. You should find $k/(2m_1)x_*^2 + gx_* = k/(2m_1)x_0^2 + gx_0$. (c) Use the six-step approach to derive the equations of motion for the bottom mass. Specialize your equations for the condition that the contact force approaches zero as $x = x_*$. (d) Combine your results to show that initial position of the top mass that causes the bottom mass to almost achieve lift-off is $x_1(0) = x_2 + \ell - (2m_1 + m_2)g/k$. Note that the answer is inversely proportional to the spring constant k: a stiff spring requires a smaller deflection than a soft spring.

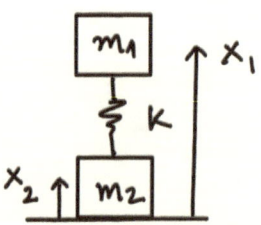

Chapter 4 Notes.

5 A Sticky Situation

An Introduction. Friction is an important and practical force to consider when creating mathematical models of complex systems. Properly including friction in the math model or in FBDs is a common source of frustration. In this section we'll focus on the friction forces in N2L and the role they play in point mass motion.

General Friction Scenarios. Four different situations can occur when a point mass is in contact with a horizontal surface:

1. The forces do not tend to move the body along the surface of contact. There is no friction force. See Figure A.

2. The applied forces tend to move the body along the surface of contact, but they are not large enough to set it in motion. The friction force that has developed can be computed by solving the equations for static equilibrium. Note, however, because there is no evidence that the friction force has reached its max value, the equation $f_{max} = \mu_s N$ cannot be used to determine the friction force. See Figure B.

3. The applied forces are such that the body is just about to slide relative to the contact surface: motion is impending. The friction force has reached its maximum value. The equations for static equilibrium and the equation $f_{max} = \mu_s N$ can be used. The friction force will have a direction that is opposite to the impending motion. See Figure C.

4. The body is sliding relative to the contact surface. The equation $f_{kin} = \mu_k N$ can be used. The friction force will have a direction that is opposite to the body's velocity relative to the contact surface. See Figure D.

⋆ 22 The Friction Illustrations.

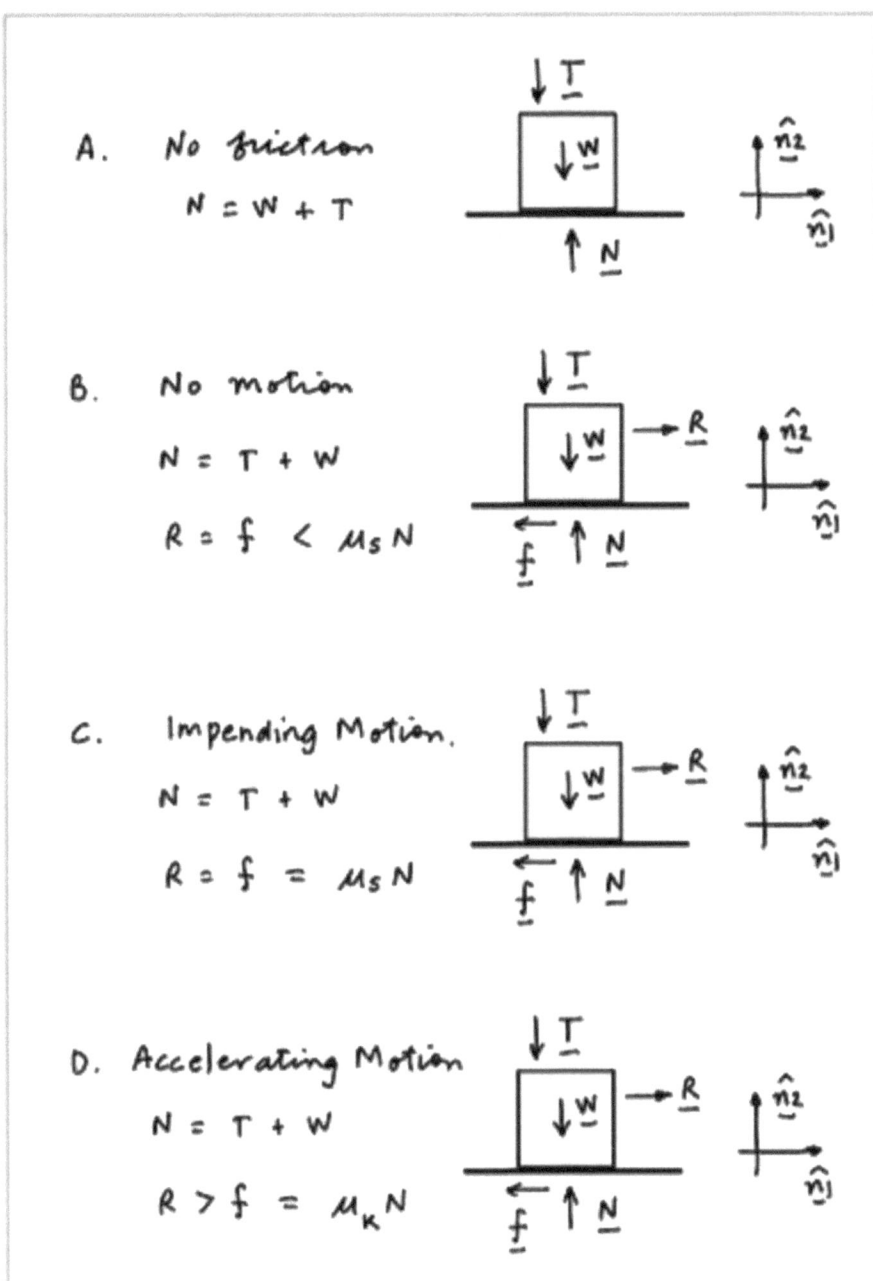

A. No friction

$N = W + T$

B. No motion

$N = T + W$

$R = f < \mu_s N$

C. Impending Motion.

$N = T + W$

$R = f = \mu_s N$

D. Accelerating Motion

$N = T + W$

$R > f = \mu_k N$

Known Impending Motion. In some problems it is stated that a point mass is on the verge of sliding relative to another body. This is commonly called impending motion. This situation involves zero relative velocity and a static friction model $f_{\max} = \mu_s N$. Consider the two cases:

1. Motion between the two bodies is impending and one is asked to determine μ_s. Here one uses static equilibrium of the forces to determine the normal N and then invokes $f_{\max} = \mu_s N$.

2. Motion between the two bodies is impending, the coefficient μ_s is given, and one is asked to determine one of the applied forces. Here one uses static equilibrium of the forces to determine the unknown force.

⋆ 23 Known Impending Motion of Stacked Blocks.

Blocks A&B are at rest and
arranged as shown. The cable
attached to A remains taut. Find
the maximum force P that can be
applied so that motion is impending.
(μs & μκ are the friction coefficients between all
surfaces.)
Steps 1,2,3,4 applied to blocks A& B.

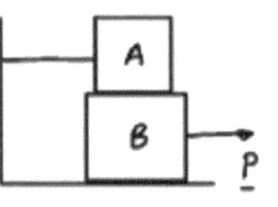

$f_A \leftarrow \quad \downarrow N_A$
$\boxed{B} \rightarrow P$

$\leftarrow$

$f_B \quad \uparrow \downarrow W_B$
N_B

$T \leftarrow \boxed{A} \rightarrow f_A$
$N_A \uparrow \downarrow W_A$

$\uparrow \hat{n}_2$
$\llcorner \rightarrow \hat{n}_1$

$B: \underline{f}_B = (-f_A - f_B + P)\ \hat{n}_1 + (N_B - N_A - W_B)\ \hat{n}_2$

$A: \underline{F}_A = (f_A - T)\ \hat{n}_1 + (N_A - W_A)\ \hat{n}_2$

step 5. Motion is impending so $\underline{a}_A = \underline{a}_B = \underline{0}$

step 6. N2L applied to A&B gives

N2L $\begin{cases} P = f_A + f_B \quad ; \quad N_B = N_A + m_B g \\ f_A = T \quad\quad ; \quad\quad N_A = m_A g \end{cases}$ $\Big\}$ $\begin{array}{c} 4 \\ EQNS \end{array}$

Friction model $\Big\{$ $f_A = \mu_s N_A \quad ; \quad f_B = \mu_s N_B$ $\Big\}$ $\begin{array}{c} 2 \\ EQNS \end{array}$

6 equations
6 unknowns $\qquad P = \mu_s m_A g + \mu_s (m_A g + m_B g)$

Example values $\qquad \begin{array}{l} m_A = 100\ kg \\ m_B = 50\ kg \end{array} \quad \mu_s = \frac{1}{4} \quad give \quad P = 613\ N$

★ 24 Known Sliding Motion of Stacked Blocks.

Blocks A&B are at rest and arranged as shown. The cable attached to A remains taut. Find the acceleration of B if $\mu_k = 0.1$ & P = 1000 N. Let $m_A = 100$ kg & $m_B = 50$ kg.

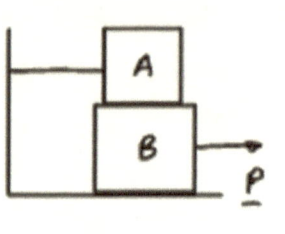

Steps 1,2,3,4 applied to blocks A & B.

$$B: \underline{f}_B = (-f_A - f_B + P)\, \hat{n}_1 + (N_B - N_A - W_B)\, \hat{n}_2$$
$$A: \underline{f}_A = (f_A - T)\, \hat{n}_1 + (N_A - W_A)\, \hat{n}_2$$

Step 5. kinematics $\quad \underline{a}_A = 0 \; ; \quad \underline{a}_B = a\, \hat{n}_1$

step 6. N2L & friction model applied to A & B gives

$$f_A = T \; ; \; N_A = m_A g$$
$$P - f_A - f_B = m_B a$$
$$N_B = N_A + m_B g$$
$$f_A = \mu_k N_A \; ; \; f_B = \mu_k N_B$$

6 Eqns
6 unks
(f_A T N_A
f_B N_B a)

Solving gives

$$m_B a = P - \mu_k m_A g - \mu_k (m_A g + m_B g)$$

so $a = 15.1 \; ^m/_{s^2}$

⋆ 25 Known Impending Motion of a Hanging Chain.

A chain of length L is released from rest with an amount C hanging over the edge of a horizontal table. Friction acts between the table surface and chain. Compute the length C so that motion is impending.

mass density

$\rho = \dfrac{mass}{length}$

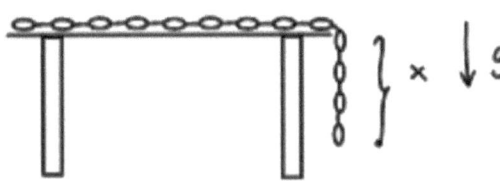

Apply 6 steps for the hanging portion.

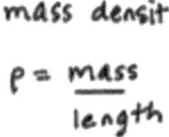

$\rho c g - T = 0$

T is an internal force

Apply 6 steps to portion remaining on the table.

$N = \rho (L-c) g$

$T - f = 0$

$f = \mu_s N$
$\quad = \mu_s \rho (L-c) g$

This gives

$T = \mu_s \rho (L-c) g$

But $T = \rho c g$ so

$\rho c g = \mu_s \rho (L-c) g$

so $\quad c = \mu_s L \,/\, (1 + \mu_s)$

⋆ 26 Known Sliding Motion of a Hanging Chain.

A chain of length L is released from rest with an amount C hanging over the edge of a horizontal table. Friction acts between the table surface and chain. Assume the length C is just enough so that sliding occurs. Develop the equations of motion. Compute the acceleration and velocity as the chain leaves the table.

$$\rho = \frac{mass}{length}$$

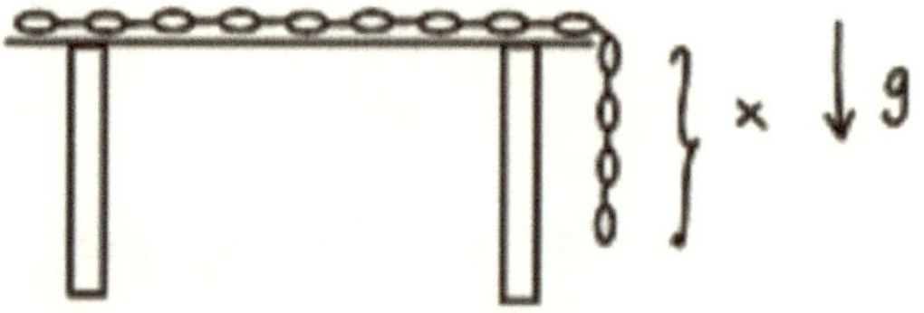

Apply 6 steps for the hanging portion.

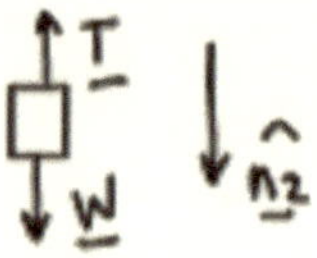

$$\rho x g - T = \rho x a$$

T is an internal force

Apply 6 steps to portion remaining on the table.

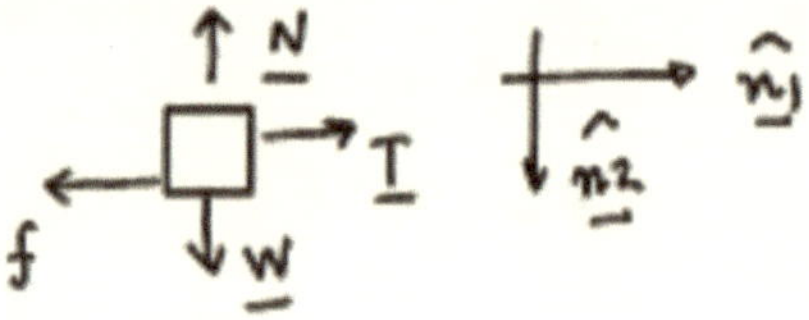

$$-N + \rho(L-x)g = 0$$

$$T - f = \rho(L-x) a$$

But $f = \mu_K N$

$$= \mu_K \rho (L-x)g$$

So

$$\rho g x - T = \rho x a$$

and

$$T - \mu_K \rho g (L-x) = \rho(L-x) a$$

Eliminating T gives $a = gx/L (1 + \mu_k) - \mu_k g$

Final acceleration: $a(x=L) = g$

Final velocity: using $a = \dfrac{dv}{dx} v$

$$\int_{c}^{L} \frac{g}{L}x \, (1 + \mu_k) \, dx - \int_{c}^{L} \mu_k g \, dx = \int_{0}^{v} v \, dv$$

$$v^2 = \frac{g}{L}(1 + \mu_k)(L^2 - c^2) - 2\mu_k g (L - c)$$

Substituting $c = \mu_s L \div (1 + \mu_s)$

$$v = \sqrt{gL (1 + 2\mu_s - \mu_k) \div (1 + \mu_s)^2}$$

⋆ 27 Another Known Sliding Motion Example.

Consider two identical blocks, each having weight W.
Let $\mu_s = \frac{1}{2}$ & $\mu_k = \frac{1}{4}$ for all contact surfaces.
Let $F = W/10$. Assume block B slips up the incline
and A slips relative to
B. Determine the
governing equations.
Compute a_A and a_B.

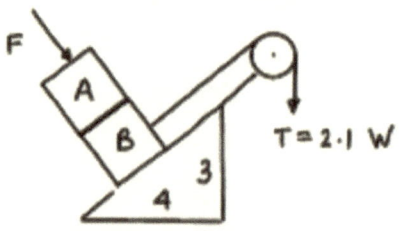

Steps 1, 2, 3, 4 for block A.

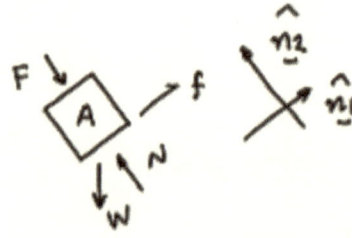

$$\underline{F}_A = -f \,\hat{n}_2 + f \,\hat{n}_1$$
$$+ N \,\hat{n}_2 - W \, 4/5 \,\hat{n}_2$$
$$- W \, 3/5 \,\hat{n}_1$$

step 5. kinematics of A. $\underline{a}_A = a_A \,\hat{n}_1$

step 6. N2L

$$m \, a_A = f - W \, 3/5$$
$$0 = N - f - W \, 4/5$$

But $f = \mu_k N = \mu_k (F + W \, 4/5)$

so

$$m \, a_A = \mu_k (F + W \, 4/5) - W \, 3/5$$

so $a_A = -3/8 \cdot g$

Steps 1, 2, 3, 4 for block B.

$$\underline{F}_B = -N\,\hat{n}_2 + L\,\hat{n}_2$$
$$- W\,{}^4/_5\,\hat{n}_2$$
$$+ T\,\hat{n}_1 - f\,\hat{n}_1$$
$$- \ell\,\hat{n}_1 - W\,{}^3/_5\,\hat{n}_1$$

step 5.   kinematics of B. $\underline{a}_B = a_B\,\hat{n}_1$

step 6.   N2L

$$m\,a_B = T - f - \ell - W\,{}^3/_5$$
$$0 = L - N - W\,{}^4/_5$$

But $f = \mu_K N = \mu_K\,(F + W\,{}^4/_5)$

and $\ell = \mu_K L = \mu_K\,(F + W\,{}^4/_5 + W\,{}^4/_5)$

so
$$m\,a_B = T - \mu_K\,(F + W\,{}^4/_5)$$
$$- \mu_K\,(F + W\,{}^4/_5 + W\,{}^4/_5)$$
$$- W\,{}^3/_5$$

so $a_B = {}^{17}/_{20} \cdot g$

Will Motion Occur? Sometimes a point mass is situated in a position of rest relative to a surface such that motion *may* occur when released. It is not evident, however, what will happen: will the point mass remain at rest relative to the other body or will it slide? Below is a method to answer such questions.

1. Solve the equations assuming static equilibrium. That is, there is no relative acceleration between the two bodies. Compute the (necessary) friction force, f_{necess}, that is consistent with this assumption. Confirm that $f_{\text{necess}} > 0$. Compare the (necessary) friction force with the maximum friction force available, $f_{\text{max}} = \mu_s N$.

   ○ If $f_{\text{necess}} \leq f_{\text{max}}$ then the original assumption was correct. The two bodies are at rest relative to each other. We are done.

   ○ If $f_{\text{necess}} > f_{\text{max}}$ then the original assumption was incorrect. Sliding takes place between the two bodies. Continue to the next step.

2. Re-solve the problem assuming that sliding takes place between the two bodies. This means $f_{\text{kin}} = \mu_k N$ and we must now use this friction model. The accelerations of the two bodies are independent.

This approach is used in the next example.

⋆ 28 A Block Worked Two Ways.

A 100 lb. block is placed at rest on a ramp and acted on by T. The force T is 90 lb. Friction acts between the contact with $\mu_s = 0.2$ & $\mu_k = 0.1$. The angle θ is 45 deg. Determine the equations of motion. Compute the block acceleration

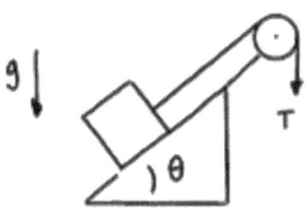

steps 1, 2, 3, 4.

$$\underline{F} = T\,\hat{n}_1 + N\,\hat{n}_2$$
$$- W\cos\theta\,\hat{n}_2$$
$$- W\sin\theta\,\hat{n}_1$$
$$- f\,\hat{n}_1$$

step 5. kinematics $\underline{r} = x\,\hat{n}_1$; $\underline{a} = a\,\hat{n}_1$

step 6.   N2L $T - f - mg\sin\theta = ma$
$$N - mg\cos\theta = 0$$

Set $a = 0$. Then $f = T - mg\sin\theta > 0$.
But $f_{max} = \mu_s N = \mu_s\,mg\cos\theta$
check $f > f_{max}$?   In other words

$T - mg\sin\theta > \mu_s\,mg\cos\theta$?

or $T > mg\,(\mu_s\cos\theta + \sin\theta) \approx 85$ lb. ?

yes. So block slides.

Thus $\quad f = \mu_k N$

so

$\quad T - \mu_k N - mg\sin\theta = ma$

$\quad T - \mu_k (mg\cos\theta) - mg\sin\theta = ma$

or

$\quad a = \dfrac{T}{m} - \mu_k g\cos\theta - g\sin\theta \approx 3.9 \text{ ft/sec.}$

Let's rework the problem assuming the block accelerates down the ramp.

steps 1, 2, 3, 4

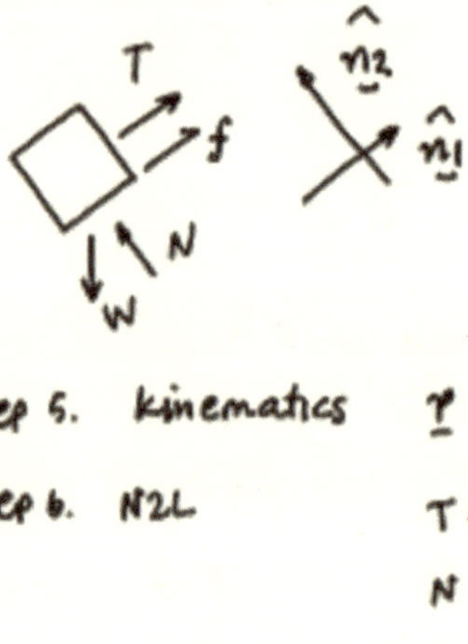

$\underline{F} = T\,\hat{n}_1 + N\,\hat{n}_2$
$\quad - W\cos\theta\,\hat{n}_2$
$\quad - W\sin\theta\,\hat{n}_1$
$\quad + f\,\hat{n}_1$

step 5. kinematics $\quad \underline{r} = -x\,\hat{n}_1 \;;\; \underline{a} = -a\,\hat{n}_1$

step 6. N2L $\qquad T + f - mg\sin\theta = -ma$

$\qquad\qquad\qquad N - mg\cos\theta = 0$

set $a = 0$. Then $\quad f = -T + mg\sin\theta < 0$

The required friction force is less than zero, so we assumed the wrong direction.

So switch
the direction
on $\underline{f}$

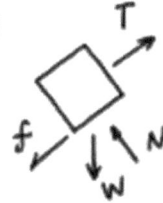

This leads to the
previous scenario.
Set $a = 0$. Then

$$f = T - mg \sin \theta$$

Is $f > f_{max}$? In other words

$$T - mg \sin \theta > \mu_s \, mg \cos \theta \; ?$$

Yes. So block slides and $f = \mu_K N$.

So
$$T - \mu_K N - mg \sin \theta = -ma$$
$$T - \mu_K (mg \cos \theta) - mg \sin \theta = -ma$$

or

$$a = \frac{-T}{m} + \mu_K g \cos \theta + g \sin \theta \approx -3.9 \; ft/sec.$$

(picked wrong direction of $\underline{a}$)

Chapter 5 Problem Set.

1. (B&J 12.31) Blocks A is at rest on a ramp that is inclined 30 deg from the horizontal. Block B rests upon block A. The coefficients of friction between block B and block A are $\mu_{sB} = 0.12$ and $\mu_{kB} = 0.10$. The coefficients of friction between block A and the ramp are $\mu_{sA} = 0.24$ and $\mu_{kA} = 0.20$. Block A has mass 10 kg whereas block B has mass 5 kg. Gravity acts along the negative vertical direction. The following questions will lead to the initial acceleration of each block. Use symbolic variables in the developments until it becomes absolutely necessary to do otherwise. (a) Follow the usual 6 steps to develop the governing equations of motion for block B. In doing this, you may assume that block B slides down relative to A. (b) Follow the usual 6 steps to develop the governing equations of motion for block A. In doing this, you may assume that block A slides down the ramp. (c) Gather the four scalar equations that govern the motion and forces of blocks A and B. What are the six unknowns in these equations? (d) The most severe case of no-motion is that $a_A = 0$, $a_B = 0$. This eliminates two of the six unknowns. Solve for the friction forces that are consistent with these assumptions. (e) Determine if the no acceleration assumptions are valid. Indeed, show that two conditional statements are revealed.

$$\tan\theta \leq \mu_{sA}? \quad \text{If yes, then block A remains at rest.}$$

$$\tan\theta \leq \mu_{sB}? \quad \text{If yes, then block B remains at rest.}$$

For the given numerical values, you should conclude that both blocks will accelerate from rest. Notice that these two conditional statements govern the four possibilities:

- Neither block accelerates from rest: $\tan\theta \leq \mu_{sA}$, $\tan\theta \leq \mu_{sB}$.

- Only block B accelerates from rest: $\tan\theta \leq \mu_{sA}$, $\tan\theta > \mu_{sB}$.

- Only block A accelerates from rest: $\tan\theta > \mu_{sA}$, $\tan\theta \leq \mu_{sB}$.

- Both blocks accelerate from rest: $\tan\theta > \mu_{sA}$, $\tan\theta > \mu_{sB}$.

(f) Because the process has confirmed that both blocks will accelerate from rest, solve for symbolic expressions of those accelerations in terms of known variables. Substitute numerical values into the symbolic expressions to compute numerical values for the accelerations. You should find $a_A = 2.78$ m/s^2 and $a_B = 4.06$ m/s^2. (g) How do your answers confirm or contradict the assumptions that B slides down relative to A and A slides down the ramp?

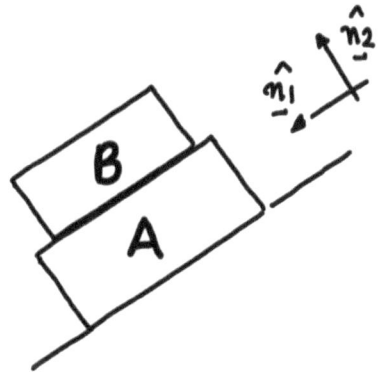

Chapter 5 Notes.

6 Point Mass Kinetics using Rotating Frames

An Introduction. It's artificial to separate this chapter of Newton's second law from the others because finding, analyzing, and solving equations of motion for point mass models is still the task. Moreover, the six-step procedure of identifying the system, drawing FBDs, establishing relevant reference frames, writing a vector representation of the forces, performing the vector kinematics up to the acceleration level, and using N2L to write the governing equations of motion is still the approach.

The difference here is that the vector forces and vector kinematics are expressed in rotating reference frames. The reason is that rotating reference frames can lead to more simple expressions that are easier to solve and more revealing.

A Motivating Example. Consider the motion of a simple pendulum, in which a point mass is tied to a constant-length string that swings from a fixed hinge in a constant gravity field. Using an angle θ to measure the pendulum position and using a fixed reference frame to express the forces and vector kinematics leads to some unpleasant-looking equations of motion.

$$-m\ell\ddot{\theta}\sin\theta - m\ell\dot{\theta}^2\cos\theta = mg - T\cos\theta \tag{99.1}$$

$$m\ell\ddot{\theta}\cos\theta - m\ell\dot{\theta}^2\sin\theta = -T\sin\theta \tag{99.2}$$

Assuming we know something about our pendulum, the unknowns in these equations are T and θ (and the time derivatives of θ). These unknowns represent the string tension and the pendulum motion variable. These equations are unpleasant because the unknowns are coupled together and there appears to be a lot of $\sin\theta$ and $\cos\theta$ present. There is a better way.

Using the same angle θ but using a rotating reference frame to express the forces and vector kinematics leads to a different set of equations of motion.

$$\ell\ddot{\theta} = -g\sin\theta \tag{99.3}$$

$$-m\ell\dot{\theta}^2 = -T + mg\cos\theta \tag{99.4}$$

Equation (99.3) is independent of the unknown string tension T, so this equation could possibly be solved for the motion variable θ. Once θ is known as a function of time, then eq. (99.4) could be used to solve for the required T.

It doesn't take a rocket scientist to see that a rotating reference frame simplified the governing equations of motion in this example. (The development of the equations for the simple pendulum is shown in the following pages.)

⋆ 29 Simple Pendulum Equations in a Fixed Reference Frame.

Consider the motion of a swinging pendulum. Suppose gravity acts downward. Derive the equations of motion.

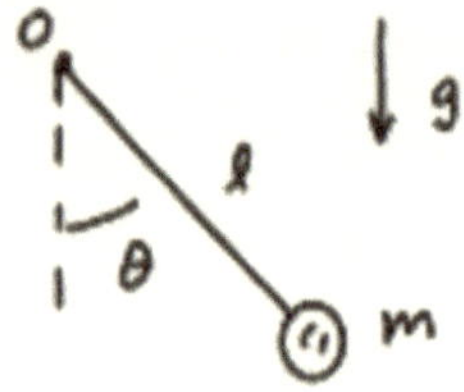

Steps 1, 2, 3, 4. (System, FBD, reference frames, vector representation of the forces.)

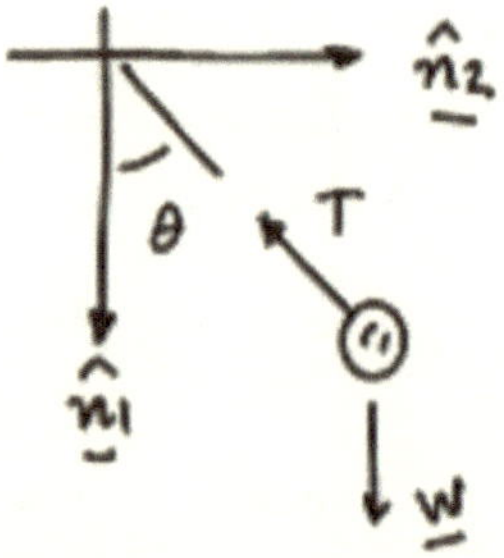

$$\underline{F} = W \, \hat{n}_1 - T \cos\theta \, \hat{n}_1$$
$$- T \sin\theta \, \hat{n}_2$$

step 5. kinematics

$$\underline{P} = \ell \cos\theta \, \hat{n}_1 + \ell \sin\theta \, \hat{n}_2$$

$$\underline{a} = (- \ell \ddot{\theta} \sin\theta - \ell \dot{\theta}^2 \cos\theta) \, \hat{n}_1$$
$$+ (\ell \ddot{\theta} \cos\theta - \ell \dot{\theta}^2 \sin\theta) \, \hat{n}_2$$

step 6. N2L

$$mg - T \cos\theta = m (- \ell \ddot{\theta} \sin\theta - \ell \dot{\theta}^2 \cos\theta)$$
$$- T \sin\theta = m (\ell \ddot{\theta} \cos\theta - \ell \dot{\theta}^2 \sin\theta)$$

⋆ 30 Simple Pendulum Equations in a Rotating Reference Frame.

Consider the motion of a swinging pendulum. Suppose gravity acts downward. Derive the equations of motion.

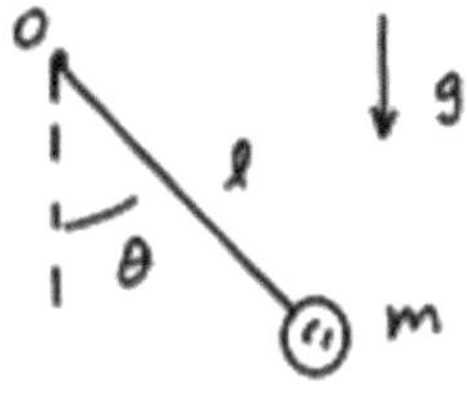

Steps 1, 2, 3, 4. (System, FBD, reference frames, vector representation of the forces.)

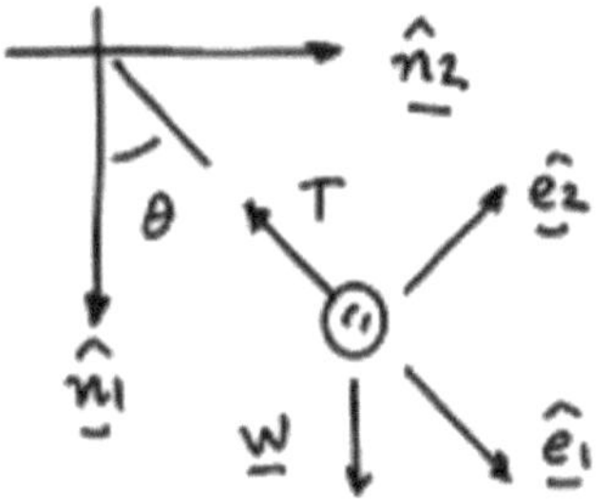

$$F = -T\,\hat{e}_1 + W\cos\theta\;\hat{e}_1$$
$$\qquad - W\sin\theta\;\hat{e}_2$$

step 5.   kinematics

$$\underline{r} = \ell\,\hat{e}_1$$

$$\underline{a} = (-\ell\dot{\theta}^2)\,\hat{e}_1 + (\ell\ddot{\theta})\,\hat{e}_2$$

step 6.   N2L

$$-mg\sin\theta = m\ell\ddot{\theta}$$
$$mg\cos\theta - T = -m\ell\dot{\theta}^2$$

A Difference Between Linear and Nonlinear Equations. The motion of a simple pendulum is governed by a constant coefficient, nonlinear, ordinary differential equation.

$$\ddot{\theta} + \frac{g}{\ell}\sin\theta = 0 \qquad (102.1)$$

One intuitively knows that the pendulum swings back and forth, and if θ starts from a small value with a small velocity, then it should not grow much larger. In such a case, because linear equations are generally easier to solve than nonlinear ones, one is tempted to replace $\sin\theta$ with θ.

$$\ddot{\theta} + \frac{g}{\ell}\theta = 0 \qquad (102.2)$$

This is a constant coefficient, linear, ordinary differential equation.

What's the difference in the solution of these two mathematical models? Plots are shown below where eqs. (102.1) and (102.2) are solved numerically using the Euler method presented earlier. The numerical solution of the nonlinear equation differs from the exact solution because of the numerical technique; the numerical solution of the linear equation differs from the exact solution because of the numerical technique and the linear assumption.

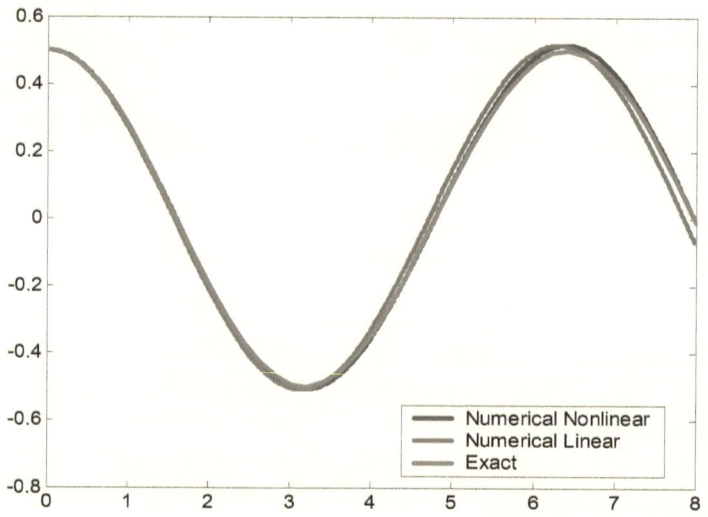

⋆ 31 The Vomit Comet.

Consider the motion of an aircraft that flies a trajectory so that students experience "weightlessness" or freefall. The speed at the top of the trajectory is v. Determine the governing equations and the instantaneous radius of curvature.

Steps 1, 2, 3, 4. (System, FBD, reference frames, vector representation of the forces.)

Free body diagram of passenger as they experience free-fall.

$$\underline{F} = - mg\, \hat{e}_1$$

step 5. kinematics $\quad \underline{p} = \rho\, \hat{e}_1$

$$\underline{v} = \dot{\rho}\, \hat{e}_1 + \rho\dot{\theta}\, \hat{e}_2 \quad = v\, \hat{e}_t$$

At top of trajectory,

$$\dot{\rho} = 0 \text{ so } \underline{v} = \rho\dot{\theta}\, \hat{e}_2 = v\, \hat{e}_2$$

$$\underline{a} = (\ddot{\rho} - \rho\dot{\theta}^2)\, \hat{e}_1 + (\rho\ddot{\theta} + 2\dot{\rho}\dot{\theta})\, \hat{e}_2$$

Setting $\dot{\rho} = \ddot{\rho} = 0$ gives

$$\underline{a} = -\rho\dot{\theta}^2\, \hat{e}_1 + \rho\ddot{\theta}\, \hat{e}_2 = \dot{v}\, \hat{e}_2 - v^2/\rho\, \hat{e}_1$$

step 6.   N2L

$$-mg = -m\,v^2/\rho \quad \text{so} \quad \rho = v^2/g$$
$$0 = m\dot{v}$$

⋆ 32 An Airplane Loop.

A pilot flies a loop in a vertical plane. Her speed is constant (v) and the radius of the loop is R. Determine the governing equations at the top B and bottom A of the loop. Compute the force of the seat acting on the pilot.

$R = 1000\,m$

B

A

Steps 1, 2, 3, 4. (System, FBD, reference frames, vector representation of the forces.)

@ top

$\hat{e}_2$ S (seat) $\underline{F} = -mg\,\hat{\underline{e}}_1 - S\,\hat{\underline{e}}_1$

$\leftarrow$ O $\uparrow \hat{e}_1$ step 5. $\underline{r} = R\,\hat{\underline{e}}_1$

$\downarrow$ W $\underline{v} = v\,\hat{\underline{e}}_2$; $\underline{a} = -v^2/R\,\hat{\underline{e}}_1$

step 6. N2L $-mg - S = -m\,v^2/R$

$S = m(v^2/R - g)$

@ bottom

$\uparrow \hat{e}_1$ $\uparrow$ S (seat) $\underline{F} = -mg\,\hat{\underline{e}}_1 + S\,\hat{\underline{e}}_1$

O $\rightarrow$ step 5. $\underline{r} = -R\,\hat{\underline{e}}_1$

$\hat{e}_2$ $\underline{v} = v\,\hat{\underline{e}}_2$; $\underline{a} = v^2/R\,\hat{\underline{e}}_1$

$\downarrow$ W

step 6. N2L $-mg + S = m\,v^2/R$

$S = m(v^2/R + g)$

⋆ **33 The Swinging Pendulum.**

A pendulum swings in a horizontal plane. Determine the equations of motion. Compute h and the string tension.

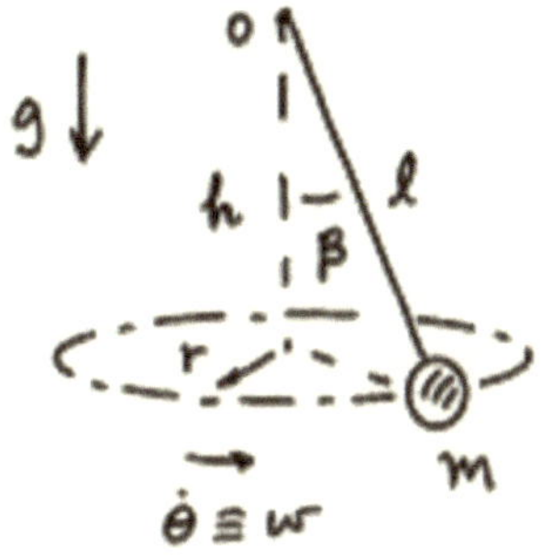

$g \downarrow$

Steps 1, 2, 3, 4. (System, FBD, reference frames, vector representation of the forces.)

$\dot{\theta} \equiv w$

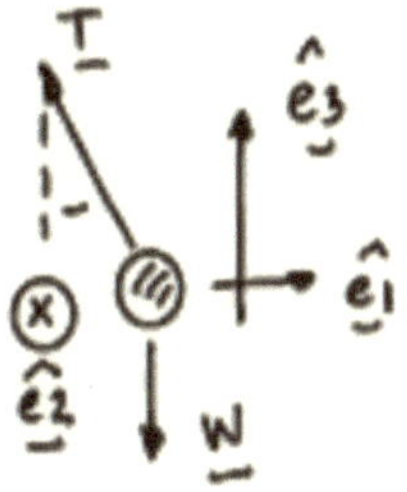

$$\underline{F} = -mg\,\hat{e}_3$$
$$-T\sin\beta\,\hat{e}_1 + T\cos\beta\,\hat{e}_3$$

side view

step 5. $\underline{r} = r\,\hat{e}_1$

$$\underline{v} = r\dot{\theta}\,\hat{e}_2$$

$$\underline{a} = -r\dot{\theta}^2\,\hat{e}_1 + r\ddot{\theta}\,\hat{e}_2$$

step 6. N2L

$$mr\ddot{\theta} = 0 \quad ; \quad -mg + T\cos\beta = 0$$
$$-mr\dot{\theta}^2 = -T\sin\beta$$

But from geometry

$$\sin\beta = r/\ell \qquad \cos\beta = h/\ell \qquad \tan\beta = r/h$$

so,
$$\tan\beta = r\dot{\theta}^2/g \quad \text{which gives} \quad h = g/w^2$$

Also

$$T\,h/\ell = mg \quad \text{so} \quad T = \frac{mg\ell}{h} = m\ell w^2$$

⋆ 34 The Block Slides Off the Dome.

A block begins from rest at the top of a circular dome. It slides along the dome if given a slight nudge. Determine the equations of motion while the block is in contact with the dome. Compute θ_* when the block loses contact.

Steps 1,2,3,4. (System, FBD, reference frames, vector representation of the forces.)

$$\underline{F} = N\,\hat{e}_1 - W\cos\theta\,\hat{e}_1 + W\sin\theta\,\hat{e}_2$$

step 5. $\quad \underline{r} = r\,\hat{e}_1$

$$\underline{v} = r\dot{\theta}\,\hat{e}_2$$

$$\underline{a} = -r\dot{\theta}^2\,\hat{e}_1 + r\ddot{\theta}\,\hat{e}_2$$

step 6. N2L

$$mg\sin\theta = mr\ddot{\theta} \quad ; \quad N - mg\cos\theta = -mr\dot{\theta}^2$$

The block loses contact if $N \to 0$ so $r\dot{\theta}^2 = g\cos\theta_*$

Using the chain rule $\quad \ddot{\theta} = \dfrac{d\dot{\theta}}{d\theta}\dfrac{d\theta}{dt} = \dfrac{d\dot{\theta}}{d\theta}\dot{\theta}$

This gives

$$r\ddot{\theta} = r\dot{\theta}\,\frac{d\dot{\theta}}{d\theta} = g\sin\theta\,.$$ Integrating gives

$$r\dot{\theta}^2 = 2g(1-\cos\theta)$$

So

$$g\cos\theta_* = 2g(1-\cos\theta_*) \quad \text{or} \quad \cos\theta_* = \frac{2}{3}$$

⋆ 35 The Block Slides Down the Bowl.

A point mass can slide
within a circular bowl.
The is friction at the
contact, and $\mu_s = 0.6$ and

$\mu_k = 0.3$. Determine the
equations of motion. Suppose
the block begins from rest when $\theta = 60$ deg. What
happens? Suppose the block begins from rest when θ
$= 90$ deg. Compute $\ddot{\theta}$ when released.

Steps 1, 2, 3, 4. (System, FBD, reference frames,
vector representation of the forces.)

$$\underline{F} = -N \, \hat{e}_1 + mg \cos\theta \, \hat{e}_1$$
$$- mg \sin\theta \, \hat{e}_2 + f \, \hat{e}_2$$

step 5.   Kinematics.

$$\underline{P} = r \, \hat{e}_1$$
$$\underline{v} = r\dot{\theta} \, \hat{e}_2$$
$$\underline{a} = -r\dot{\theta}^2 \, \hat{e}_1 + r\ddot{\theta} \, \hat{e}_2$$

step 6.   N2L

$$-N + mg \cos\theta = -mr\dot{\theta}^2$$
$$f - mg \sin\theta = mr\ddot{\theta}$$

The point mass begins from rest when $\theta = 60$ deg
Does motion occur?

Suppose not. $\underline{a} = \underline{0}$ which gives $f = mg \sin \theta$.

But $f_{max} = \mu_s N = \mu_s mg \cos \theta$

So if $f > f_{max}$ then point mass slides.

If $f < f_{max}$ then point mass does not slide.

$mg \sin \theta > \mu_s mg \cos \theta$?

$\tan \theta > \mu_s$?

$\tan (60 \text{ deg}) > 0.6$?   Yes. Motion occurs

Suppose the point mass begins from rest when $\theta = 90$.

Then, because $\dot{\theta} = 0$ and $\cos (90) = 0$, $N = 0$.

Since $N = 0$, then $f = 0$.

So $- mg \sin (90 \text{ deg}) = mr \ddot{\theta}$

or $- g/r = \ddot{\theta}$

⋆ 36 Satellite in Elliptic Orbit.

A satellite travels in an elliptical orbit about the Earth. The semi major axis a is known; the semi minor axis b is known; and the eccentricity $e = h/a$ is known. At the instant shown, the satellite speed v is known.

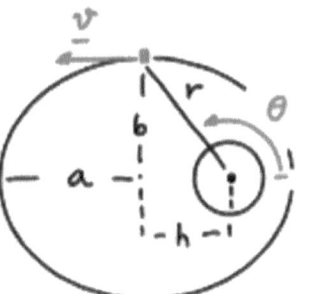

Note that the attractive force due to gravity is $F = mg \dfrac{R^2}{r^2}$ where R is the Earth radius.

Determine the equations of motion. Compute $\dot{r}$, $\dot{\theta}$, $\ddot{r}$ and $\ddot{\theta}$ in terms of known values.

Steps 1,2,3,4. (System, FBD, reference frames, vector representation of the forces.)

$$\underline{F} = -mg \frac{R^2}{r^2} \hat{e}_1$$

step 5. kinematics

$$\underline{P} = r\,\hat{e}_1$$
$$\underline{v} = \dot{r}\,\hat{e}_1 + r\dot{\theta}\,\hat{e}_2$$
$$\underline{a} = (\ddot{r} - r\dot{\theta}^2)\,\hat{e}_1$$
$$\qquad + (r\ddot{\theta} + 2\dot{r}\dot{\theta})\,\hat{e}_2$$

step 6. N2L

$$-mg\frac{R^2}{r^2} = m(\ddot{r} - r\dot{\theta}^2) \;;\qquad 0 = r\ddot{\theta} + 2\dot{r}\dot{\theta}$$

But $\underline{v} = -v \, \hat{n}_1 = -v \, (\cos\theta \, \hat{e}_1 - \sin\theta \, \hat{e}_2)$

So $-v\cos\theta = \dot{r}$ and $v\sin\theta = r\dot{\theta}$

But $\theta = \pi - \beta$ and $\cos\theta = -\cos\beta$ and $\sin\theta = \sin\beta$.

$\cos\beta = h/r = ea \div \sqrt{e^2 a^2 + b^2}$

$\sin\beta = b/r = b \div \sqrt{e^2 a^2 + b^2}$

so

$\dot{r} = v \, ea \div \sqrt{e^2 a^2 + b^2}$

$\dot{\theta} = vb \div (e^2 a^2 + b^2)$

$\ddot{\theta} = -\dfrac{2\dot{r}\dot{\theta}}{r} = -2 \dfrac{v^2 e a b}{(e^2 a^2 + b^2)^2}$

$\ddot{r} = -g\dfrac{R^2}{r^2} + r\dot{\theta}^2$ { substitute and compute }

Chapter 6 Problem Set.

1. (M&K 3.54) It is observed that an aircraft is undergoing a bank maneuver in a horizontal plane. The force on the body due to lift is perpendicular to the wing surface. (a) Follow the usual 6 steps to develop the governing equations of motion. (b) Compute the required bank angle for an aircraft flying at a speed of 400 mi/hr and making a circular turn with radius $r = 2$ mi. Answer: $\theta = 45.32$ deg.

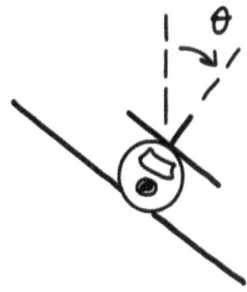

2. A point mass can slide along a straight wire of length ℓ, which rotates about the end point o at a constant rate $\dot\theta$ in a horizontal plane. Gravity acts downward. A linear spring with constant k connects the point mass to the end of the wire as shown, and the spring is unstretched when the mass has zero radial displacement. All surfaces are smooth and there is no appreciable friction between the point mass and wire. (a) Follow the usual 6 steps to develop the governing equations of motion.

3. Consider a stiff, straight wire that spins in a horizontal plane about one end. The wire spins at a known constant rate $\dot\theta$. A bead with known mass m is threaded on the spinning wire and held at rest relative to the wire at the known location r_* when it is suddenly released. Friction acts between the bead and the wire, with known static and kinetic friction coefficients μ_s and μ_k, respectively. Gravity acts along the negative vertical direction. For simplicity, take the gravitational value to equal $g = 10 \text{ m/s}^2$. Answer the series of questions below, which are organized to determine the initial radial acceleration $\ddot r$ of the bead. (a) Follow the usual 6 steps to develop the governing equations of motion. (b) Recall that the bead is placed at

rest relative to the wire at the location r_* and then released. Derive the inequality expression that tests whether the bead will undergo an initial radial acceleration $\ddot{r}$. (c) For $m = 3$ kg, $\dot{\theta} = 2$ rad/sec, $r_* = 1/2$ meter, $\mu_s = 0.18$, and $\mu_k = 0.1$, show that the bead will undergo an initial radial acceleration. (d) Using the above parameter values, compute the initial radial acceleration. (e) Solve for a symbolic expression that relates the current radial speed $\dot{r}$ to the current radial position r, the original radial position r_*, and other given parameters or values. In doing so, make the assumption that the friction force only depends on the normal force $N = mg$.

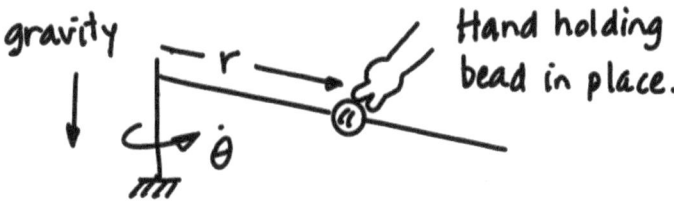

4. Consider the motion of a point mass that is being pulled by a thin, non-stretching string that remains taut. The movement of mass m is confined to the horizontal table. All surfaces are smooth and there is no appreciable friction between the point mass and table surface. Gravity acts downward.

(a) Follow the usual 6 steps to develop the governing equations of motion.

(b) Now suppose the movement is such that $\dot{r}$ is constant. Show that the corresponding force is $T = c/r^3$ where c is a constant.

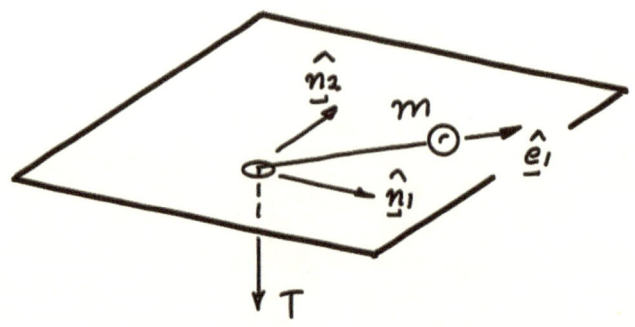

5. A circular disk shown in the figure below spins in a horizontal plane. A mass m kg rests on the disk at a distance r inches from the axis of rotation. Friction acts between the point mass and surface with known static and kinetic friction coefficients μ_s and μ_k, respectively. Gravity acts along the negative vertical direction. The disk starts from rest with a known constant angular acceleration $\ddot{\theta} = \alpha$ rad/sec^2. Experimental observations reveal that the particle will begin to slip at some point in time. (a) Follow the usual 6 steps to demonstrate that the governing equations of motion up to the point that the particle will slip are as follows:

$$f_1 = mr\alpha^2 t^2 \quad ; \quad f_2 = mr\alpha \quad ; \quad N = mg$$

Here, f_1 and f_2 are friction forces along the radial and tangential directions, respectively. (b) The particle will slip when the magnitude of the friction force equals $f_{\max} = \mu_s N$. Show that the time when this occurs can be computed from $t = (1/\alpha)\sqrt[4]{(\mu_s g/r + \alpha)(\mu_s g/r - \alpha)}$.

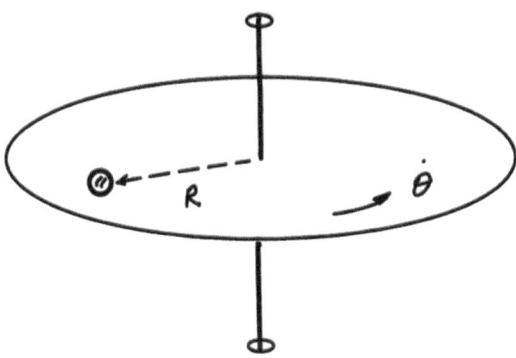

6. (M&K 3.72) A common amusement ride looks like a large pirate ship that swings back and forth about a hinge at o. A motor drives the ship so that the swinging motion is described as $\theta(t) = A\sin(\omega t)$, where A is the amplitude and ω is the frequency. Let $A = 10$ m and let $\omega = 1$ rad/sec. (a) Follow the usual 6 steps to develop the governing equations of motion for a passenger sitting in the middle of the ship. In doing so, assume the seat and seatbelt provides reaction forces on the passenger that act radially and tangentially. (b) Use the governing equations to write an expression for the normal force that the seat provides to the rider. (c) Plot the normal force as a function of time for 10 seconds. Also plot θ as a function of time for that time interval. (d) Repeat all derivations, computations, etc. for a rider sitting at the end of the ship.

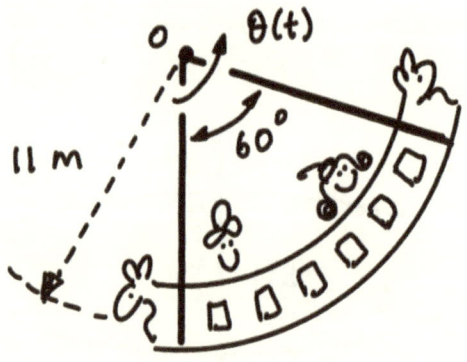

7. Two equal 2 kg masses are rigidly connected and separated from each other by 30 deg. Friction acts between mass A and the surface of a circular half-pipe, but not between mass B and the surface. The connecting bar follows the circular half-pipe so that the connection force on each mass is parallel to the local tangent. The masses are released from rest when $\theta = 30$ deg. The coefficients of friction are $\mu_s = 3/4$ and $\mu_k = 1/\sqrt{3}$. Gravity acts downward, and for simplicity let $g = 10$ m/s². The half-pipe has radius 1 m. (a) Fully and clearly develop all governing equations for each point mass. (b) Demonstrate whether the system accelerates from rest or not. (c) Compute the initial tangential acceleration of mass A and the force in the rigid connection between the two masses at the instant when the system is released. (d) Is the connecting bar in tension or compression?

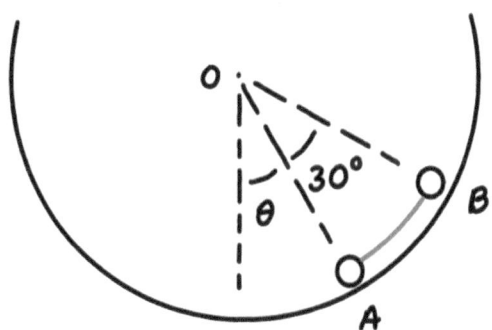

8. (M&K 3.96) A helicopter transitions from rest in a hover mode to straight and level horizontal flight with a known acceleration a_h. A payload hangs from the helicopter and is free to oscillate in the plane. (a) Follow the usual 6 steps to develop the governing equations of motion for the payload. (b) The payload is initially at rest relative to the helicopter. Use this truth together with the chain rule to find an expression that relates the instantaneous $\dot{\theta}$ to the instantaneous θ. (c) Use the results obtained thus far to relate the cable tension with the swing angle θ.

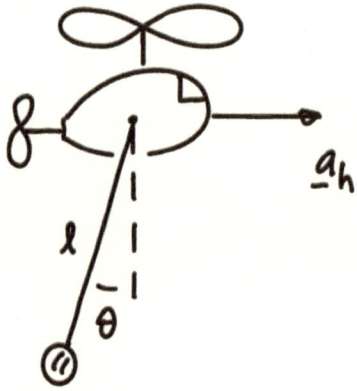

Chapter 6 Notes.

7 Point Mass Angular Momentum

An Introduction. Angular momentum is a useful concept for understanding motion that encircles a reference point. This type of motion happens in the orbital movement of heavenly bodies and the tumbling movement of rigid bodies. Angular momentum is a defined quantity involving position and velocity relative to a reference point. In this section we'll introduce angular momentum and briefly look at its role in point mass motion.

The Definition. The angular momentum vector of a point mass around a fixed point o is defined as $\boldsymbol{h}_o \equiv \boldsymbol{p} \times m\dot{\boldsymbol{p}} = \boldsymbol{p} \times m\boldsymbol{v}$. The vectors $\boldsymbol{p}$ and $\boldsymbol{v}$ are the inertial position and velocity vectors of the point mass relative to point o. The reference point o is critical in the definition of angular momentum: one can not simply say 'the angular momentum vector', one must mention the reference location.

The angular momentum vector is created from the cross product of the position and (mass-weighted) velocity vector, or the position and translational momentum vector. The expression for the angular momentum vector can be constructed once the vector kinematics have been developed: write the inertial position vector, compute the inertial velocity vector, take the cross product, done.

To illustrate, recall the simple pendulum. The inertial position vector along a rotating reference frame is $\boldsymbol{p} = \ell\hat{\boldsymbol{e}}_1$. The inertial velocity vector is $\boldsymbol{v} = \ell\dot{\theta}\hat{\boldsymbol{e}}_2$. The angular momentum vector about the hinge can now be computed.

$$\boldsymbol{h}_o = \boldsymbol{p} \times m\boldsymbol{v} = m\ell^2\dot{\theta}\hat{\boldsymbol{e}}_3. \qquad (122.1)$$

The position and velocity vector lie in a plane, and the angular momentum vector is perpendicular to that plane. So although we've tried to keep our dynamics discussion strictly planar, this presentation (or representation) of angular momentum forces us to include another direction, even if the motion vectors never leave the original plane.

⋆ 37 A Point Mass Launched off a Table.

A point mass is launched horizontally off a table with
speed v_0. Gravity acts downward. Determine the angular
momentum vector of the point
mass relative to the corner.

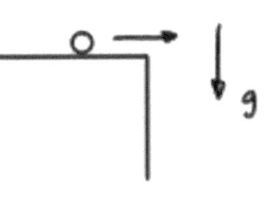

Steps 1,2,3,4. (System, FBD, reference frames,
vector representation of the forces.)

$\hat{n}_1$

step 5. Kinematics.

$\underline{p} = x \, \hat{\underline{n}}_1 + y \, \hat{\underline{n}}_2$

$\hat{\underline{n}}_2$

$\underline{v} = \dot{x} \, \hat{\underline{n}}_1 + \dot{y} \, \hat{\underline{n}}_2$

$\underline{W}$

$\underline{h}_O = \underline{p} \times m \underline{v}$

$\underline{h}_O = (x \, \hat{\underline{n}}_1 + y \, \hat{\underline{n}}_2) \times m (\dot{x} \, \hat{\underline{n}}_1 + \dot{y} \, \hat{\underline{n}}_2)$

$= (x \dot{y} - y \dot{x}) m \, \hat{\underline{n}}_3$

But $\dot{x} = v_0$ for all time, so $x = v_0 t$.

And $\dot{y} = v_y = g t$ so $y = \frac{1}{2} g t^2$

Then $\underline{h}_O = (v_0 t \, g t - \frac{1}{2} g t^2 \, v_0) m \, \hat{\underline{n}}_3$

$= m \, \frac{1}{2} v_0 g t^2 \, \hat{\underline{n}}_3$

The Time Derivative of h_o. The angular momentum vector of a point mass around a fixed point o is defined as $h_o \equiv p \times m\dot{p} = p \times mv$. Let's investigate the time derivative of h_o.

$$\dot{h}_o = p \times ma = p \times f \qquad (124.1)$$

Newton's second law has been used to introduce the external forces acting on the point mass. Note that the vector $\dot{h}_o$ can equal zero under two conditions: if the sum of the external forces is zero, $f = 0$; or if the position vector and the sum of the external forces are collinear, $p \times f = 0$. If either of these are true, then $\dot{h}_o = 0$, which means the angular momentum vector is conserved: the magnitude and direction never change.

★ 38 $\dot{h}_o$ for the Point Mass Launched off a Table.

A point mass is launched horizontally off a table with
speed v_0. Gravity acts downward. Is the angular momentum
vector relative to the corner
conserved?

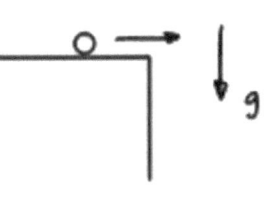

Steps 1, 2, 3, 4. (System, FBD, reference frames,
vector representation of the forces.)

$\hat{n}_1$

forces: $\underline{f} = mg\,\hat{n}_2$

○ (x,y)

step 5. Kinematics.

$\hat{n}_2$

$\underline{r} = x\,\hat{n}_1 + y\,\hat{n}_2$

$\underline{W}$

$\underline{h}_o = \underline{r} \times \underline{f} = -mg\,x\,\hat{n}_3$

But $\dot{x} = v_0$ for all time, so $x = v_0 t$

so

$\underline{\dot{h}}_o = mg\,v_0\,t\,\hat{n}_3$

$\underline{\dot{h}}_o \neq 0$

The Mutual Attraction of Two Bodies. Newton's fourth law says that the magnitude of the attractive force between two bodies is $f = GMm/r^2$, where G is the universal gravitational constant, M and m are the two mass values, and r is the distance between them. This force is directed along the line connecting the two bodies, so vectorially one may write $\boldsymbol{f} = (GMm/r^3)\boldsymbol{r}$.

One can use this result to find the equations that govern the *relative* motion between two bodies. Consider the Moon (m) orbiting the Earth (M) where each is treated as a point mass. The motion of the Moon relative to the Earth is found by following our usual steps of identifying the system, drawing FBDs, establishing relevant reference frames, writing a vector representation of the forces, performing the vector kinematics up to the acceleration level, and using N2L. Using a rotating reference frame with polar coordinates leads to the following.

$$\ddot{r} - r\dot{\theta}^2 = -\frac{\mu}{r^2} \qquad ; \qquad 0 = r\ddot{\theta} + 2\dot{r}\dot{\theta} \qquad (126.1)$$

The parameter $\mu = G(M + m)$ is a constant.

These equations are thoroughly presented in the following pages.

⋆ 39 Development of the Relative 2-Body Equations.

Develop the relative two-body equations of motion.

System, FBD, reference frames, vector representation of the forces, etc. for the two bodies.

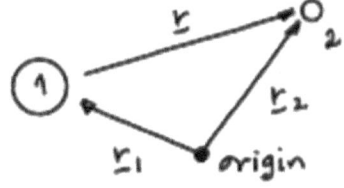

$$M \underline{a_1} = \underline{f} = G\frac{Mm}{r^3}\underline{r}$$

$$m \underline{a_2} = -\underline{f} = -G\frac{Mm}{r^3}\underline{r}$$

Define relative kinematic variables.

$$\underline{r} = \underline{r_2} - \underline{r_1}$$
$$\underline{v} = \underline{v_2} - \underline{v_1}$$
$$\underline{a} = \underline{a_2} - \underline{a_1}$$

so

$$\underline{a} = \underline{a_2} - \underline{a_1} = -G\frac{M}{r^3}\underline{r} - G\frac{m}{r^3}\underline{r}$$

$$\underline{a} = -G(M+m)\frac{\underline{r}}{r^3} = -\frac{\mu}{r^3}\underline{r}$$

$$\underline{r} = r\,\hat{e_1}$$
$$\underline{v} = \dot{r}\,\hat{e_1} + r\dot{\theta}\,\hat{e_2}$$
$$\underline{a} = (\ddot{r} - r\dot{\theta}^2)\,\hat{e_1} + (r\ddot{\theta} + 2\dot{r}\dot{\theta})\,\hat{e_2}$$

so

$$(\ddot{r} - r\dot{\theta}^2)\,\hat{e_1} + (r\ddot{\theta} + 2\dot{r}\dot{\theta})\,\hat{e_2} = -\frac{\mu}{r^2}\,\hat{e_1}$$

Angular Momentum of the Relative 2-Body Problem. In celestial mechanics, the equations of motion for one body relative to another are found using N2L and N4L.

$$\ddot{r} - r\dot{\theta}^2 = -\frac{\mu}{r^2} \quad ; \quad 0 = r\ddot{\theta} + 2\dot{r}\dot{\theta} \tag{128.1}$$

The second equation is a statement of the relative angular momentum. Let's see.

The angular momentum vector of the Moon relative to the Earth is $\boldsymbol{h} = \boldsymbol{p} \times \boldsymbol{v} = r^2\dot{\theta}\hat{\boldsymbol{e}}_3$. This vector is conserved, i.e., has constant magnitude and direction, because the relative position vector and attractive force vector are collinear.

$$\dot{\boldsymbol{h}} = \boldsymbol{p} \times \boldsymbol{f} = r\hat{\boldsymbol{e}}_1 \times \left(-\mu/r^2\right)\hat{\boldsymbol{e}}_1 = \boldsymbol{0} \tag{128.2}$$

Let's investigate the conservation of angular momentum in light of the equations of motion.

$$\boldsymbol{h} = \boldsymbol{p} \times \boldsymbol{v} = r\hat{\boldsymbol{e}}_1 \times (\dot{r}\hat{\boldsymbol{e}}_1 + r\dot{\theta}\hat{\boldsymbol{e}}_2) = r^2\dot{\theta}\hat{\boldsymbol{e}}_3 = h\hat{\boldsymbol{e}}_3 \tag{128.3}$$

$$\dot{h} = r^2\ddot{\theta} + 2r\dot{r}\theta = r(r\ddot{\theta} + 2\dot{r}\dot{\theta}) = 0 \text{ (from N2L)} \tag{128.4}$$

So, the second equation of motion from N2L [the right-most expression in (128.1)] is a statement of the conservation of angular momentum.

The other equation [the left-most expression in (128.1)] is also a statement of something, but let's first investigate a neat change of variables.

Presto Chango for the Relative 2-Body Problem. Let's use a new co-ordinate and the rules of calculus to simplify one of the governing equations for the relative two-body problem.

1. First, recall the angular momentum variable, $h = r^2 \dot{\theta}$ and note that h is constant.

2. Second, consider a new *dependent* coordinate s so that $sr = 1$. Always.

3. Next, consider a change of the *independent* variable from t to θ. Here, the chain rule of calculus is handy.

$$\frac{d(*)}{dt} = \frac{d(*)}{d\theta}\frac{d\theta}{dt} = \frac{d(*)}{d\theta}\dot{\theta} \qquad \text{so, for example} \qquad \dot{s} = s'\dot{\theta} \qquad (129.1)$$

4. Next, investigate the time derivative of $sr = 1$ to find a replacement for $\dot{r}$.

$$0 = \dot{s}r + s\dot{r} = s'\dot{\theta}r + s\dot{r} = s'h/r + s\dot{r} = hss' + s\dot{r} \Rightarrow 0 = hs' + \dot{r} \quad (129.2)$$

5. Next, use the above items to find a replacement for $\ddot{r}$.

$$\ddot{r} = d(-hs')/dt = d(-hs')/d\theta\, \dot{\theta} = -hs''\dot{\theta} = -h^2s^2s'' \qquad (129.3)$$

The above items can be used to rewrite one of the governing equations for the relative two-body problem.

$$\ddot{r} - r\dot{\theta}^2 = -\frac{\mu}{r^2} \quad \rightarrow \quad -s^2h^2\left(s'' + s - \frac{\mu}{h^2}\right) = 0 \quad \Rightarrow \quad s'' + s = \frac{\mu}{h^2} \quad (129.4)$$

Look what we've done. The left-most equation is a coupled, nonlinear, homogeneous ordinary differential equation. These are generally difficult to solve. The right-most is an uncoupled, linear, nonhomogeneous ordinary differential equation. These are relatively easy to solve.

⋆ 40 A Presto Chango Exercise.

Carry out the steps on the previous page to show

$$\text{this } \ddot{r} - r\dot{\theta}^2 = -\frac{\mu}{r^2}$$

$$\text{becomes this } -s^2 h^2 \left(s'' + s - \frac{\mu}{h^2} \right) = 0$$

$$\text{which implies this } s'' + s = \frac{\mu}{h^2}$$

Energy of the Relative 2-Body Problem. The equations of motion for one body relative to another are becoming familiar.

$$\ddot{r} - r\dot{\theta}^2 = -\frac{\mu}{r^2} \quad ; \quad 0 = r\ddot{\theta} + 2\dot{r}\dot{\theta} \tag{131.1}$$

We've already seen that the second equation is a statement of the relative angular momentum. The first equation can be swapped for something else using a change in the dependent and independent variables.

$$s'' + s = \frac{\mu}{h^2} \tag{131.2}$$

We can take a step toward solving this equation by noting that s'' is an acceleration and s is a position; s' would be the velocity.

$$s'' = \frac{d(s')}{d\theta} = \frac{d(s')}{ds}\frac{ds}{d\theta} = \frac{d(s')}{ds}s' \tag{131.3}$$

Using this in (131.2) and integrating is familiar.

$$s'\,ds' = -s\,ds + \frac{\mu}{h^2}\,ds \quad \rightarrow \quad s'^2 + s^2 = 2\frac{\mu}{h^2}s - E \tag{131.4}$$

We won't prove it here and we'll have more to say about this subject later, but the right-most part of (131.4) is the energy expression for the relative two-body problem: the two terms on the left are the relative kinetic energy, the first term on the right is the potential energy of the inverse square law, and E is the constant of integration. Recall that all of this flowed from the first equation of motion.

In summary, the differential equations of motion for the relative two-body problem are statements of the relative motion energy and relative motion angular momentum. Both energy and angular momentum are constant.

There's quite a bit more to the relative two-body problem, but we'll stop here.

(One More Integration Reveals Conic Sections.) One of the equations that governs relative two-body motion is $\ddot{r} - r\dot{\theta}^2 = -\mu/r^2$, which can be replaced with a more suitable equation.

$$s'' + s = \mu/h^2 \qquad (132.1)$$

This replacement is a linear, second order, ordinary differential equation for s as a function of θ. The angle θ is called the *true anomaly* by the way.

Two integrations of (132.1) are needed to compute $s(\theta)$. We already performed one to uncover the energy equation.

$$s'^2 + s^2 = 2\frac{\mu}{h^2}s - E \qquad (132.2)$$

The remaining integration, from (132.2) to $s(\theta)$, is more elaborate than the integration of (132.1) to (132.2). It involves algebraic and calculus techniques, like completing the square and a change of integration variables, that are commonly covered when learning integral calculus methods. Here are the details.

1. First, recalling that $s' \equiv \mathrm{d}s/\mathrm{d}\theta$, separate the independent and dependent variables in (132.2).

$$s'^2 + s^2 = 2\frac{\mu}{h^2}s - E \qquad \rightarrow \qquad \frac{\mathrm{d}s}{\sqrt{-s^2 + 2s/p - E}} = \mathrm{d}\theta \qquad (132.3)$$

Here we've introduced the constant parameter $p = h^2/\mu$. This is commonly called the *semi-latus rectum*.

2. Complete the square of the radicand.

$$-s^2 + 2s/p - E = (1/p^2 - E) - (s - 1/p)^2 \qquad (132.4)$$

3. Introduce a new, temporary variable $w = s - 1/p$ and a new, temporary constant $P^2 = 1/p^2 - E$ within the radicand. Notice $\mathrm{d}w = \mathrm{d}s$.

$$(1/p^2 - E) - (s - 1/p)^2 = P^2 - w^2 \qquad (132.5)$$

At this point eq. (132.3) has been transformed.

$$\frac{\mathrm{d}s}{\sqrt{-s^2 + 2s/p - E}} = \mathrm{d}\theta \qquad \rightarrow \qquad \frac{\mathrm{d}w}{\sqrt{P^2 - w^2}} = \mathrm{d}\theta \qquad (132.6)$$

4. Introduce another new, temporary variable $w = P \sin \eta$ and note that $dw = P \cos \eta \, d\eta$. This transforms (132.6) into something trivial.

$$d\eta = d\theta \qquad \rightarrow \qquad \eta = \theta + c \qquad (132.7)$$

5. All that's left now is to unravel the temporary transformations.

$$\eta = \theta + c \quad \rightarrow \quad w = P \sin(\theta + c) \quad \rightarrow \quad s = 1/p + P \sin(\theta + c)$$

$$\rightarrow \quad s = 1/p + P \cos c \sin \theta + P \sin c \cos \theta$$

$$\rightarrow \quad s = 1/p + (s_0 - 1/p) \cos \theta \qquad (132.8)$$

A particular constant of integration was selected in the final step.

Although we've finally computed $s(\theta)$, we recall that the original motion coordinate was r and not s.

$$sr = 1 \quad \rightarrow \quad r = \frac{1}{1/p + (1/r_0 - 1/p) \cos \theta} \quad \rightarrow \quad r = \frac{p}{1 + (p/r_0 - 1) \cos \theta}$$

$$\rightarrow \quad r = \frac{p}{1 + e \cos \theta} \qquad (132.9)$$

A new constant e, called the *eccentricity*, has been introduced in the final step.

One may recall from their high school analytic geometry course that (132.9) is the equation of a conic section (more specifically, an ellipse) in polar coordinates with its origin at the focus. Thus the relative orbital motion of one body to another is an ellipse as evidenced by the fact that Newton's laws of motion using polar coordinates are satisfied by $r = p/(1 + e \cos \theta)$.

⋆ 41 Relative 2-Body Motion.

A course in celestial mechanics teaches that orbits are described by conic sections, i.e., ellipses, parabolas, and hyperbolas. This is seen in the equation that relates the radial position to the angular position.

$$r = p/\left(1 + e\cos\theta\right) \tag{134.1}$$

One can graph this curve using a polar plot, as shown below. The open circle marks the starting point of the orbit. The ellipse parameters p and e are shown, along with the *semi major axis a* and *semi minor axis b*.

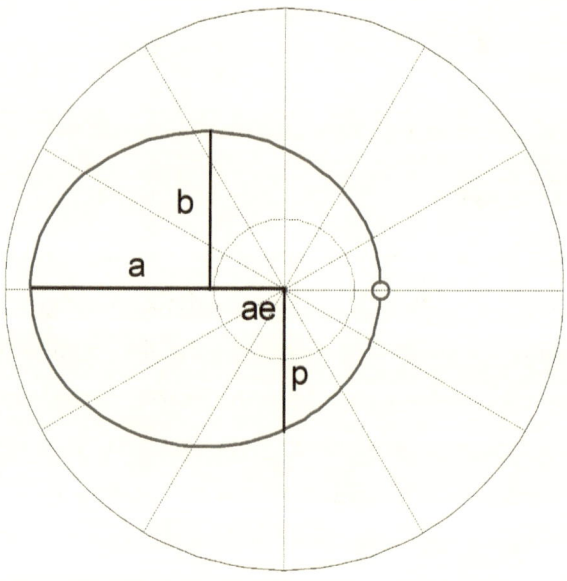

Chapter 7 Problem Set.

1. You're considering the thought that Newton got it wrong. You're supposing that the force of gravity is not inversely proportional to distance squared, but simply proportional to distance. That is,

> A student said they'd pursue
>
> A different point of view
>
> That Newton was wrong
>
> For the force is but strong
>
> As the distance 'tween the two
>
> (J.E.H)

(a) Solve for the relative two-body equations of motion in terms of polar coordinates r and θ. You may assume that motion takes place in the plane, and the only force acting on each point mass is the attraction of gravity, which acts along the line connecting the two bodies. The magnitude of the gravitational force is $F = Gm_1m_2r$, where r is the distance between the two bodies. (b) The answer above is two scalar equations. Write every mathematical adjective that truthfully describes each of these two equations. (c) Show that one of the equations is a statement that the relative angular momentum is constant. (d) Your differential equations above are written using polar coordinates r and θ. What are the governing equations in terms of Cartesian coordinates x and y? (e) Write every mathematical adjective that truthfully describes each of these new equations. (f) Is the relative angular momentum still constant? Show through computation with the new Cartesian coordinates that it is, or is not, constant. (g) Consider your two differential equations from question d. Demonstrate that $x(t) = x_0 \cos(kt)$; $y(t) = (\dot{y}_0/k)\sin(kt)$ will satisfy the differential equations. Here, k, x_0, and $\dot{y}_0$ are constants, and t is time.

Chapter 7 Notes.

8 Rigid Body Kinematics

An Introduction. A rigid body model of a system differs significantly from a point mass model in that a rigid body model accommodates rotational motion. This means that a rigid body model can undergo translational and rotational motions. Investigations of the translational motion reveals familiar kinematic expressions. Investigations of the rotational motion, however, reveals something new: the angular velocity vector.

Position and Attitude. When confined to a plane, a total of three coordinates are needed to completely specify the movement of a rigid body. Two coordinates are needed to pinpoint the position of a point on the body relative to a fixed origin, and one coordinate is needed to pinpoint the rotation, or attitude, or orientation of the body relative to a fixed direction.

For convenience, consider a point called the rigid body mass center.[7] The position of the mass center relative to an inertial origin can be written using Cartesian or polar coordinates, along fixed or rotating reference frame directions.

$$\boldsymbol{p}_c = x\,\hat{\boldsymbol{n}}_1 + y\,\hat{\boldsymbol{n}}_2 = r\cos\phi\,\hat{\boldsymbol{n}}_1 + r\sin\phi\,\hat{\boldsymbol{n}}_2 = r\,\hat{\boldsymbol{e}}_1 = \sqrt{x^2+y^2}\,\hat{\boldsymbol{e}}_1 \qquad (139.1)$$

The two position coordinates could be (x,y) or (r,ϕ).

The rotation or attitude or orientation of the rigid body can be described by considering the angular displacement of a single fixed line within the rigid body relative to a fixed inertial direction. Let $\hat{\boldsymbol{b}}_1$ and $\hat{\boldsymbol{b}}_2$ be orthogonal fixed lines in a rigid body and let $\hat{\boldsymbol{n}}_1$ and $\hat{\boldsymbol{n}}_2$ be inertially fixed orthogonal directions. Then either $\hat{\boldsymbol{b}}_1$ or $\hat{\boldsymbol{b}}_2$ could serve as the single fixed line in the rigid body.

$$\hat{\boldsymbol{b}}_1 = \cos\theta\,\hat{\boldsymbol{n}}_1 + \sin\theta\,\hat{\boldsymbol{n}}_2 \quad ; \quad \hat{\boldsymbol{b}}_2 = -\sin\theta\,\hat{\boldsymbol{n}}_1 + \cos\theta\,\hat{\boldsymbol{n}}_2 \qquad (139.2)$$

In either case, the one attitude coordinate is θ. The $\hat{\boldsymbol{b}}_1$ and $\hat{\boldsymbol{b}}_2$ vectors define a body-fixed reference frame b^+ that translates and rotates with the rigid body.

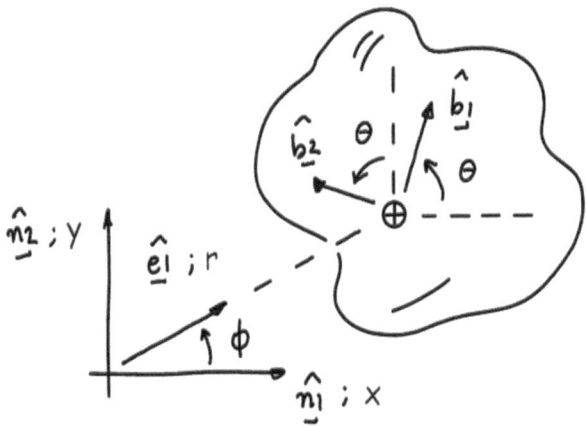

[7]This is a point about which the mass is equally distributed.

Body-Fixed Reference Frames. It is convenient to use $\hat{b}_1$ and $\hat{b}_2$ orthogonal lines fixed in a rigid body to describe the rotation or attitude or orientation of a rigid body relative to inertial directions. A single angle θ is an appropriate attitude coordinate for planar rotations.

$$\hat{b}_1 = \cos\theta\,\hat{n}_1 + \sin\theta\,\hat{n}_2 \quad ; \quad \hat{b}_2 = -\sin\theta\,\hat{n}_1 + \cos\theta\,\hat{n}_2 \qquad (140.1)$$

It is worth emphasizing something that was mentioned on the previous page: an important new concept here in discussing rigid body attitude is the use of a *body-fixed reference frame*, which is constructed of the $\hat{b}_1$ and $\hat{b}_2$ unit vectors. The body-fixed reference frame is denoted as b^+, and it translates and rotates with the rigid body. It does whatever the body does. The concept of a body-fixed reference frame is used in more general 3-dimensional rotational dynamics, too.

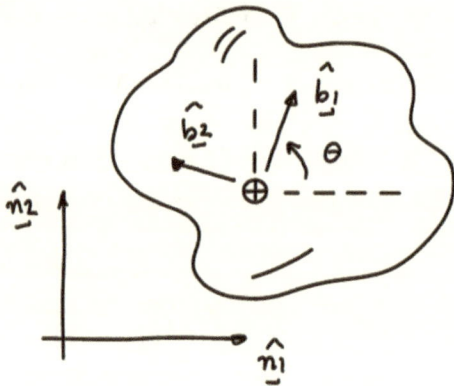

Velocities: Specifically, What is Angular Velocity?   Again, consider the position of a rigid body mass center using Cartesian coordinates along inertially fixed directions.

$$\boldsymbol{p}_c = x\,\hat{\boldsymbol{n}}_1 + y\,\hat{\boldsymbol{n}}_2 \tag{141.1}$$

The translational velocity vector of the mass center is the straightforward time derivative of the position vector.

$$\boldsymbol{v}_c = \dot{x}\,\hat{\boldsymbol{n}}_1 + \dot{y}\,\hat{\boldsymbol{n}}_2 = v_x\,\hat{\boldsymbol{n}}_1 + v_y\,\hat{\boldsymbol{n}}_2 \tag{141.2}$$

Now consider the attitude of the rigid body described by the angular displacement of a fixed line within the rigid body relative to a fixed inertial direction. Either $\hat{\boldsymbol{b}}_1$ or $\hat{\boldsymbol{b}}_2$ could serve as the single fixed line.

$$\hat{\boldsymbol{b}}_1 = \cos\theta\,\hat{\boldsymbol{n}}_1 + \sin\theta\,\hat{\boldsymbol{n}}_2 \quad ; \quad \hat{\boldsymbol{b}}_2 = -\sin\theta\,\hat{\boldsymbol{n}}_1 + \cos\theta\,\hat{\boldsymbol{n}}_2 \tag{141.3}$$

Of course, the attitude of these vectors relative to the inertially fixed directions could change with time if θ changes with time. To understand the velocity associated with attitude or rotational motion, consider the time derivative of either $\hat{\boldsymbol{b}}_1$ or $\hat{\boldsymbol{b}}_2$.

$$\mathrm{d}/\mathrm{dt}(\hat{\boldsymbol{b}}_1) = \dot{\theta}\hat{\boldsymbol{b}}_2 \quad ; \quad \mathrm{d}/\mathrm{dt}(\hat{\boldsymbol{b}}_2) = -\dot{\theta}\hat{\boldsymbol{b}}_1 \tag{141.4}$$

We first saw these expressions when we discussed point mass kinematics using rotating reference frames. Note that $\hat{\boldsymbol{b}}_2$ could be replaced with the cross product of unit vectors $\hat{\boldsymbol{b}}_1$ and $\hat{\boldsymbol{b}}_3$. A similar truth holds for $\hat{\boldsymbol{b}}_1$.

$$\hat{\boldsymbol{b}}_2 = \hat{\boldsymbol{b}}_3 \times \hat{\boldsymbol{b}}_1 \quad ; \quad \hat{\boldsymbol{b}}_1 = -\hat{\boldsymbol{b}}_3 \times \hat{\boldsymbol{b}}_2 \tag{141.5}$$

Thus, eqs. (141.4) can be rewritten.

$$\mathrm{d}/\mathrm{dt}(\hat{\boldsymbol{b}}_1) = \dot{\theta}\hat{\boldsymbol{b}}_3 \times \hat{\boldsymbol{b}}_1 \quad ; \quad \mathrm{d}/\mathrm{dt}(\hat{\boldsymbol{b}}_2) = \dot{\theta}\hat{\boldsymbol{b}}_3 \times \hat{\boldsymbol{b}}_2 \tag{141.6}$$

This expression inspires us to define $\dot{\theta}\hat{\boldsymbol{b}}_3$ as *the rigid body angular velocity vector*, $\boldsymbol{\omega} = \omega\,\hat{\boldsymbol{b}}_3 = \dot{\theta}\,\hat{\boldsymbol{b}}_3$.

$$\mathrm{d}/\mathrm{dt}(\hat{\boldsymbol{b}}_1) = \boldsymbol{\omega} \times \hat{\boldsymbol{b}}_1 \quad ; \quad \mathrm{d}/\mathrm{dt}(\hat{\boldsymbol{b}}_2) = \boldsymbol{\omega} \times \hat{\boldsymbol{b}}_2 \tag{141.7}$$

Translational and Rotational Acceleration. The translational acceleration vector of the mass center is the time derivative of the translational velocity
vector. This is straightforward for Cartesian coordinates along inertially fixed
directions.

$$a_c = \ddot{x}\,\hat{n}_1 + \ddot{y}\,\hat{n}_2 = a_x\,\hat{n}_1 + a_y\,\hat{n}_2 \qquad (142.1)$$

The rigid body angular acceleration vector for planar motion is the time
derivative of the angular velocity vector.

$$\boldsymbol{\alpha} = \alpha\,\hat{b}_3 = \dot{\omega}\,\hat{b}_3 = \ddot{\theta}\,\hat{b}_3 \qquad (142.2)$$

★ **42 Looping Maneuver of a Rigid Body Airplane.**

An aircraft performs a loop maneuver with constant speed v_0 and constant pitch rate $w = \dot{\theta}$. Show that the revolution rate and rotation rate are equal.

The translational kinematics of the mass center are carried out.

Let $\phi \equiv$ the revolution angle.

$$\underline{r}_c = R\,\hat{e}_1$$

$$\underline{v}_c = R\dot{\phi}\,\hat{e}_2 = v_0\,\hat{e}_2$$

The revolution rate: $\dot{\phi} = \dfrac{v_0}{R}$. The revolution period equals $t = 2\pi R / v_0$.

The rotational kinematics about the mass center are carried out. For a loop maneuver, the rotational period equals the revolution period.

So, $\dfrac{2\pi R}{v_0} = \dfrac{2\pi}{\dot{\theta}}$ or $\dot{\theta} = \dfrac{v_0}{R}$

The rotation rate & revolution rates are equal.

$\theta \equiv$ rotation angle

(An Angular Position Vector?) The attitude of a rigid body was introduced by relating the angular displacement of a fixed line in the rigid body to a fixed inertial direction. We've used $\hat{b}_1$ and $\hat{b}_2$ which compose a body-fixed reference frame as examples.

$$\hat{b}_1 = \cos\theta\,\hat{n}_1 + \sin\theta\,\hat{n}_2 \quad ; \quad \hat{b}_2 = -\sin\theta\,\hat{n}_1 + \cos\theta\,\hat{n}_2 \qquad (144.1)$$

The time derivative of these vectors led to something called the rigid body angular velocity vector.

$$\boldsymbol{\omega} = \omega\,\hat{b}_3 = \dot{\theta}\,\hat{b}_3 \qquad (144.2)$$

It is tempting to use this result to introduce something called an angular position vector.

That is, if $\boldsymbol{\omega} = \omega\,\hat{b}_3 = \dot{\theta}\,\hat{b}_3$ is an angular velocity vector, then isn't $\boldsymbol{\theta} = \theta\,\hat{b}_3$ an angular position vector?

In this special case, yes, $\boldsymbol{\theta} = \theta\,\hat{b}_3$ can be considered an angular position vector. But this is only valid for planar rigid body motions, it is *not* valid for general 3-dimensional rotational motions: generally speaking, an angular position vector is nonsensical. So, because conventional aerospace systems, like aircraft and spacecraft, undergo 3-dimensional rotational motions, a concept like an angular position vector has very limited use.

Alternatively, using relative angular displacements between fixed lines in a rigid body and fixed lines in inertial space (like (144.1) above) is always valid. And since angular velocity was built from that idea, the angular velocity vector is always valid. So, the angular velocity vector is always valid but the angular position vector is not. Weird.

Position and Velocity of Two Points on the Same Body. One can write a kinematic relationship between two points on the same rigid body. First, consider a position relationship.

$$\boldsymbol{p}_\diamond = \boldsymbol{p}_c + \boldsymbol{p}_{\diamond/c} \tag{145.1}$$

Here, $\diamond$ denotes an arbitrary point on a body and c denotes the mass center. This expression states that the inertial position vector of $\diamond$ equals the inertial position vector of the mass center plus the position vector of $\diamond$ relative to the mass center. That is, $\boldsymbol{p}_{\diamond/c}$ is the relative position vector.

The time derivative of (145.1) gives a velocity expression.

$$\boldsymbol{v}_\diamond = \boldsymbol{v}_c + \boldsymbol{v}_{\diamond/c} \tag{145.2}$$

Now, suppose $\boldsymbol{p}_{\diamond/c}$ is written with vector components in a body-fixed reference frame, $\boldsymbol{p}_{\diamond/c} = p_1\,\hat{\boldsymbol{b}}_1 + p_2\,\hat{\boldsymbol{b}}_2$. The time derivative of $\boldsymbol{p}_{\diamond/c}$ is the velocity $\boldsymbol{v}_{\diamond/c}$. In computing $\boldsymbol{v}_{\diamond/c}$ we note that the body-fixed vectors $\hat{\boldsymbol{b}}_1$ and $\hat{\boldsymbol{b}}_2$ change because the rigid body can generally rotate, but the measures p_1 and p_2 do not change because the body is rigid.

$$\boldsymbol{v}_{\diamond/c} = p_1\,\mathrm{d}(\hat{\boldsymbol{b}}_1)/\mathrm{d}t + p_2\,\mathrm{d}(\hat{\boldsymbol{b}}_2)/\mathrm{d}t = p_1\,\boldsymbol{\omega} \times \hat{\boldsymbol{b}}_1 + p_2\,\boldsymbol{\omega} \times \hat{\boldsymbol{b}}_2 = \boldsymbol{\omega} \times (p_1\hat{\boldsymbol{b}}_1 + p_2\hat{\boldsymbol{b}}_2)$$

$$\rightarrow \boldsymbol{v}_{\diamond/c} = \boldsymbol{\omega} \times \boldsymbol{p}_{\diamond/c} \tag{145.3}$$

This can be used in (145.2).

$$\boldsymbol{v}_\diamond = \boldsymbol{v}_c + \boldsymbol{\omega} \times \boldsymbol{p}_{\diamond/c} \tag{145.4}$$

Truthfully, $\diamond$ and c are just two points on the same rigid body, so eq. (145.4) holds for arbitrary points on the same rigid body.

$$v_2 = v_1 + \boldsymbol{\omega} \times \boldsymbol{p}_{2/1} \tag{145.5}$$

This equation says that the inertial velocity of a point on a rigid body relative to another point on the same body is composed of two parts. One is a translational part, v_1, and the other is a rotational part, $\boldsymbol{\omega} \times \boldsymbol{p}_{2/1}$.

⋆ 43 Velocities of a Sliding Rod.

A rigid bar slides along a wall such that the ends stay in contact. End B slides with known velocity v_B. Compute the velocity of A and the angular velocity vector for the instant shown.

The velocity relationship for ends A & B.

$$\underline{v}_A = \underline{v}_B + \underline{\omega} \times \underline{r}_{A/B}$$

$$\underline{v}_A = v_A \, \hat{n}_2$$

$$\underline{v}_B = -v_b \, \hat{n}_1$$

$$\underline{\omega} = -\omega \, \hat{n}_3$$

$$\underline{r}_{A/B} = -4 \, \hat{n}_1 + 3 \, \hat{n}_2$$

$$v_A \, \hat{n}_2 = -v_B \, \hat{n}_1 + (-\omega \, \hat{n}_3) \times (-4 \, \hat{n}_1 + 3 \, \hat{n}_2)$$

$$= -v_B \, \hat{n}_1 + 4\omega \, \hat{n}_2 + 3\omega \, \hat{n}_1$$

So $v_A = 4\omega$ and $0 = -v_B + 3\omega$

Solving $\omega = v_B/3$ and $v_A = \dfrac{4}{3} v_B$

⋆ 44 Revolution and Rotation of the Moon.

Consider the motion of our moon around the Earth. The point N always faces Earth. Show that the orbital rate equals the rotation rate.

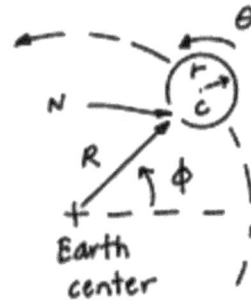
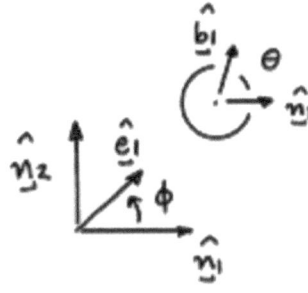

ϕ is the revolution angle ; θ is the rotation angle

$$\underline{w} \text{ Moon} = w \, \hat{\underline{n}_3} = w \, \hat{\underline{e}_3} = w \, \hat{\underline{b}_3}$$

Point N moves on a circular path of radius R.

So, $\underline{v}_N = R \dot{\phi} \, \hat{\underline{e}_2}$

Point C moves on a circular path of radius R+r.

So, $\underline{v}_C = (R+r) \dot{\phi} \, \hat{\underline{e}_2}$

But $\underline{v}_N = \underline{v}_C + \underline{w} \times \underline{r}_{N/C}$

$$R\dot{\phi} \hat{\underline{e}_2} = (R+r)\dot{\phi} \, \hat{\underline{e}_2} + \dot{\theta} \, \hat{\underline{e}_3} \times (-r \, \hat{\underline{e}_1})$$

$$= (R+r)\dot{\phi} \, \hat{\underline{e}_2} - r\dot{\theta} \, \hat{\underline{e}_2}$$

So $\dot{\phi} = \dot{\theta}$, or orbital rate = rotation rate

Acceleration of Two Points on the Same Rigid Body. The velocity relationship between two points on a rigid body involves the rigid body angular velocity and the distance between the two points.

$$v_2 = v_1 + \omega \times p_{2/1} \tag{148.1}$$

The acceleration relationship between the same two points involves the rigid body angular acceleration, the rigid body angular velocity, and the distance between the two points.

$$a_2 = a_1 + \alpha \times p_{2/1} + \omega \times v_{2/1} \tag{148.2}$$

or

$$a_2 = a_1 + \alpha \times p_{2/1} + \omega \times (\omega \times p_{2/1}) \tag{148.3}$$

⋆ 45 Accelerations of a Sliding Rod.

A rigid bar slides in a circular pipe. The ends stay in contact. End B slides with known acceleration a_B. The bar begins from rest. Compute the acceleration of end A for the instant shown. Compute the angular acceleration $\underline{\alpha}$.

The acceleration relationship is

$$\underline{a}_A = \underline{a}_B + \underline{\alpha} \times \underline{r}_{A/B}$$
$$+ \underline{w} \times (\underline{w} \times \underline{r}_{A/B})$$

$$\underline{a}_B = -a_B \,\hat{n_1}$$

$$\underline{a}_A = a_A \frac{1}{\sqrt{2}} \hat{n_1} + a_A \frac{1}{\sqrt{2}} \hat{n_2}$$

$$\underline{\alpha} = -\alpha \,\hat{n_3} \quad ; \quad \underline{w} = \underline{0} \quad (\text{begins from rest.})$$

$$\underline{r}_{A/B} = L \cos 22.5° \,\hat{n_2}$$
$$- L \sin 22.5 \,\hat{n_1}$$

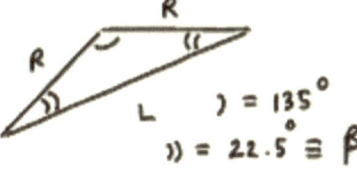

$) = 135°$

$)) = 22.5° \equiv \beta$

Substitute into above

$$\frac{a_A}{\sqrt{2}} = -a_B + \alpha L \cos\beta \quad ; \qquad \frac{a_A}{\sqrt{2}} = \alpha L \sin\beta$$

$$a_A = \frac{-\sqrt{2}\,a_B \sin\beta}{\sin\beta - \cos\beta} \quad ; \quad \alpha = \frac{1}{L} \frac{-a_B}{\sin\beta - \cos\beta}$$

Rotational Motion Only. Expressions that relate the kinematics for two points on the same rigid body have been developed.

$$v_2 = v_1 + \omega \times p_{2/1} \tag{150.1}$$

$$a_2 = a_1 + \alpha \times p_{2/1} + \omega \times (\omega \times p_{2/1}) \tag{150.2}$$

Consider the special case that point 1 is fixed in space, which means the rigid body is free to rotate about a fixed point but cannot translate. The rigid body can undergo pure rotational motion only. Let p represent the position of a point of the rigid body measured from the fixed point. The velocity and acceleration vectors are denoted as v and a. The kinematic expression are a simplified version of the above.

$$v = \omega \times p \tag{150.3}$$

$$a = \alpha \times p + \omega \times v \tag{150.4}$$

⋆ 46 A Rotating Disk.

A circular disk rotates about its mass center c. At the instant shown, the acceleration of the edge point

□ is $\underline{a} = -3\,\hat{b_2} - 4\,\hat{b_1}\ \dfrac{in}{sec^2}$.

Compute $\underline{\alpha}$ & $\underline{\omega}$.

The kinematics for rotational motion are

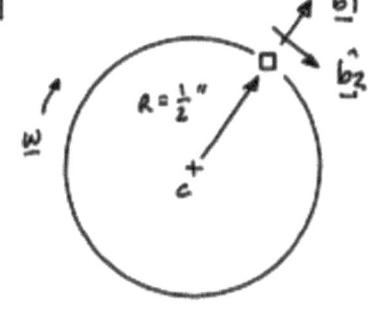

$\underline{\rho} = R\,\hat{b_1}$

$\underline{v} = \underline{\omega} \times \underline{\rho}$

$= \omega\,\hat{b_3} \times R\,\hat{b_1}$

$= R\omega\,\hat{b_2}$

$\underline{a} = \underline{\alpha} \times \underline{\rho} + \underline{\omega} \times (\underline{\omega} \times \underline{\rho})$

$= \alpha\,\hat{b_3} \times R\,\hat{b_1} + \omega\,\hat{b_3} \times R\omega\,\hat{b_2}$

$= R\alpha\,\hat{b_2} - R\omega^2\,\hat{b_1}$

But $\underline{a} = -3\,\hat{b_2} - 4\,\hat{b_1}$ in/sec^2

so $\quad \alpha = -\dfrac{3}{R}$; $\underline{\alpha} = -6\,\hat{b_3}$ rad/sec^2

$-R\omega^2 = -4$; $\underline{\omega} = \sqrt{8}\,\hat{b_3}$ rad/sec

Kinematics of Roll Without Slip. When two rigid bodies with smooth edges are in contact with one another, it is possible for the two bodies to roll without slip against each other. Roll without slip means that the instantaneous contact point of the two bodies share the same velocity.

A classic example of this is a wheel of radius R rolling without slip on a stationary flat surface. The surface is stationary so all points of the surface have zero velocity. Thus, the point on the wheel that is instantaneously in contact with the surface has zero velocity. This fact results in a relationship between the translational and angular velocities of the wheel. Let's see.

$$p_\diamond = p_c + p_{\diamond/c} \tag{152.1}$$

Here, $\diamond$ denotes the point on the wheel that is in contact with the flat surface and c denotes the mass center.

$$v_\diamond = v_c + \omega \times p_{\diamond/c} \tag{152.2}$$

For a straight horizontal surface, $v_c = \dot{x}\,\hat{n}_1$ and $\omega = -\dot{\theta}\,\hat{n}_3$. The position vector of the contact point relative to the mass center is $p_{\diamond/c} = -R\hat{n}_2$. Because the wheel rolls without slip, $v_\diamond = 0$. When all of this is used in (152.2), an unsurprising kinematic relationship is established.

$$\dot{x} = R\dot{\theta} \qquad \rightarrow \qquad x = R\theta \tag{152.3}$$

For roll without slip, the translational motion of the mass center is tied to the rotational motion of the wheel.

Roll Without Slip Accelerations. When a wheel rolls without slip on a stationary flat surface, the point on the wheel that is instantaneously in contact with the surface has zero velocity. This leads to a relationship between the translational motion of the mass center and the rotational motion of the wheel.

$$v_\diamond = 0 \quad \rightarrow \quad \dot{x} = R\dot{\theta} \quad \rightarrow \quad x = R\theta \tag{153.1}$$

This implies $\ddot{x} = R\ddot{\theta}$, and this plays a role in computing the acceleration on the wheel that is instantaneously in contact with the surface.

Consider the acceleration relationship between two points on the same rigid body, where $\diamond$ denotes the point on the wheel that is in contact with the flat surface and c denotes the mass center.

$$a_\diamond = a_c + \alpha \times p_{\diamond/c} + \omega \times (\omega \times p_{\diamond/c}) \tag{153.2}$$

The goal is to compute $a_\diamond$ and this can be easily done by computing the right side term-by-term.

$$a_c = \ddot{x}\,\hat{n}_1 \tag{153.3}$$

$$\alpha \times p_{\diamond/c} = (-\ddot{\theta}\hat{n}_3) \times (-R\hat{n}_2) = -R\ddot{\theta}\hat{n}_1 \tag{153.4}$$

$$\omega \times (\omega \times p_{\diamond/c}) = (-\dot{\theta}\hat{n}_3) \times ((-\dot{\theta}\hat{n}_3) \times (-R\hat{n}_2)) = R\dot{\theta}^2\hat{n}_2 \tag{153.5}$$

Only one part survives when all of this is used in (153.2).

$$a_\diamond = \ddot{x}\,\hat{n}_1 - R\ddot{\theta}\hat{n}_1 + R\dot{\theta}^2\hat{n}_2 = R\dot{\theta}^2\hat{n}_2 \tag{153.6}$$

If a wheel rolls without slip on a flat surface then the acceleration vector on the wheel that is instantaneously in contact with the surface is solely pointed upwards, towards the mass center. Also, the translational motion of the mass center is still tied to the rotational motion of the wheel.

$$\ddot{x} = R\ddot{\theta} \quad \rightarrow \quad a_c = R\alpha \tag{153.7}$$

⋆ 47 A Disk Rolls on a Dome.

A circular disk rolls without slip on the circular arc. At the instant shown, the angular velocity is $\underline{\omega} = \omega \, \hat{n}_3$ and the angular acceleration is $\underline{\alpha} = -\alpha \, \hat{n}_3$. Compute the acceleration vector of the mass center and of the contact point in terms of $\underline{\omega}$ and $\underline{\alpha}$.

Note: Point c moves on a circular path of radius $R+r$.

$$\underline{r}_c = (R+r)\,\hat{e}_1$$

$$\underline{v}_c = (R+r)\,\dot{\phi}\,\hat{e}_2$$

$c \equiv$ mass ctr $d \equiv$ contact pt.

$$\underline{a}_c = -(R+r)\,\dot{\phi}^2\,\hat{e}_1 + (R+r)\,\ddot{\phi}\,\hat{e}_2$$

But $\underline{v}_d = 0$ (rolls w/out slip)

Also, $\underline{\omega}_{disk} = \omega\,\hat{n}_3 = \omega\,\hat{e}_3 = \omega\,\hat{b}_3$ where b^+ is a body-fixed frame.

b^+ is body-fixed

Now,

$$\underline{v}_c = \underline{v}_d + \underline{\omega} \times \underline{r}_{c/d}$$

$$= 0 + \underline{\omega}\,\hat{e}_3 \times r\,\hat{e}_1$$

$$= r\omega\,\hat{e}_2$$

So $(R+r)\,\dot{\phi} = r\omega$ or $\dot{\phi} = \dfrac{r\omega}{(R+r)}$

Taking a time derivative of this result.

$$\ddot{\phi} = \frac{r \, d\omega / dt}{(R+r)} = \frac{-r\alpha}{(R+r)}$$

So

$$\underline{a}_c = -(R+r)\dot{\phi}^2 \, \hat{e_1} + (R+r)\ddot{\phi} \, \hat{e_2}$$

$$= -\frac{r^2\omega^2}{(R+r)} \, \hat{e_1} - r\alpha \, \hat{e_2}$$

Now,

$$\underline{a}_d = \underline{a}_c + \underline{\alpha} \times \underline{r}_{d/c} + \underline{\omega} \times (\underline{\omega} \times \underline{r}_{d/c})$$

$$= -\frac{r^2\omega^2}{(R+r)} \, \hat{e_1} - r\alpha \, \hat{e_2}$$

$$+ (-\alpha \, \hat{e_3}) \times (-r \, \hat{e_1}) + \omega \, \hat{e_3} \times (-r\omega \, \hat{e_2})$$

$$= \left(-\frac{r^2\omega^2}{(R+r)} + r\omega^2 \right) \hat{e_1} - (r\alpha - r\alpha) \, \hat{e_2}$$

So

$$\underline{a}_d = \frac{Rr\omega^2}{(R+r)} \, \hat{e_1}$$

Chapter 8 Problem Set.

1. (B&J 15.40) Rod AB moves such that end A slides along the flat floor while a portion smoothly slides along the contact at C. At the instant shown, the velocity of end A is 5 in/sec to the left and the acceleration of end A is 1 in/sec^2 to the right. (a) It is important to note that the velocity vector of the point on the rod that is in contact at C must lie solely along the rod. Why is this true? (b) Use vector kinematics for two points on the same rigid body to compute the angular velocity vector of the rod at this instant. (c) Use vector kinematics for two points on the same rigid body to compute the angular acceleration vector of the rod at this instant.

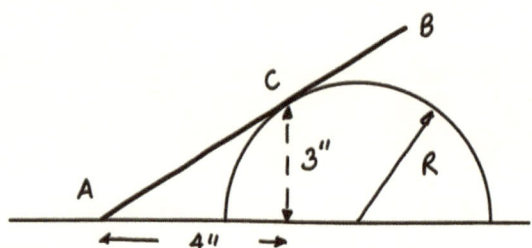

2. (M&K 5.104) A slender rod is leaning against a smooth semicircular surface. End A is given a known velocity v to the right. Throughout the
   motion, the end A remains in contact with the floor and the rod touches
   and slides along the semicircular surface. (a) Establish a reference frame
   that you will use to express all kinematic vectors. (b) Show that the speed
   of end B equals the speed of end A when the midpoint comes into contact
   with the semicircle.

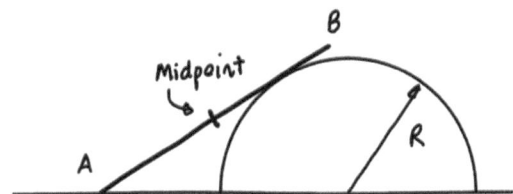

3. A circular disk rolls without slip in a larger circular bowl. At the instant shown the angular velocity of the disk is $\boldsymbol{\omega} = \omega\,\hat{n}_3$ and the angular acceleration is $\boldsymbol{\alpha} = -\alpha\,\hat{n}_3$. Compute the acceleration vector of the disk mass center and the acceleration vector of the contact point between the disk and bowl. Your answers should be in terms of ω and α.

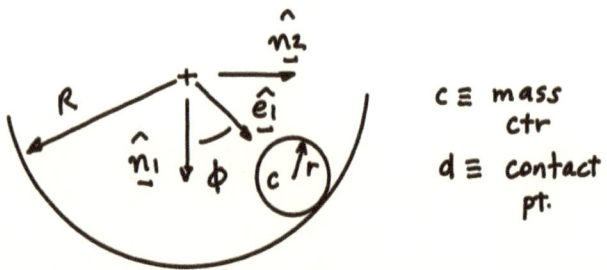

$C \equiv$ mass ctr

$d \equiv$ contact pt.

4. At the instant shown, the angular velocity of the disk is given as w in the clockwise direction. Its angular acceleration is given as α in the counterclockwise direction. The disk has radius R and rolls without slip. (a) Establish a reference frame that you will use to express all kinematic vectors. (b) Use vector kinematics for two points on the same rigid body to compute the location(s) on the disk (relative to the contact point) that has (have) zero acceleration at this instant. Your answers must be in terms of the given parameters.

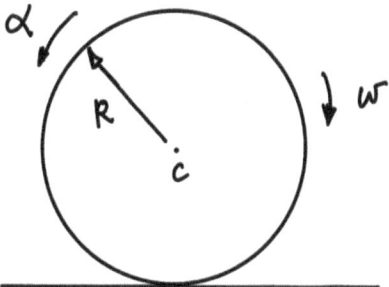

5. The landing gear of an aircraft is being retracted. The translational velocity of hinge O is known to equal v to the left. The landing strut is being
retracted with a known constant angular velocity $\dot{\theta}$ relative to the n^+ reference frame in the counterclockwise direction. Also, the tire is spinning
at a known angular velocity w, also in the counterclockwise direction, relative to the n^+ frame. The tire has radius R and the strut length from O
to C is L. (a) Compute the velocity vector of point P on the tire, which
at the instant shown is in line with the strut. Express your answer in the
n^+ frame. Let $\theta = 45$ deg.

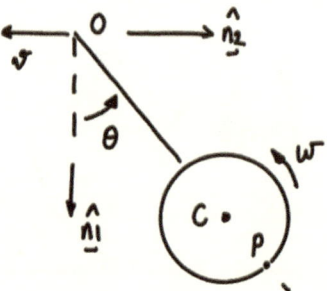

6. (B&J 11.70) Consider the two bar mechanism shown. Points O and C are inertially-fixed points along the same horizontal line. Bar OA is driven at a known angular velocity in the counterclockwise direction. The pin on bar OA moves in a slot of bar CB, which causes bar CB to rotate. (a) Show that the angular velocity of bar CB is one-half that of bar OA, regardless of the angle θ. That is, $\omega_{CB} = \frac{1}{2}\omega_{OA}$.

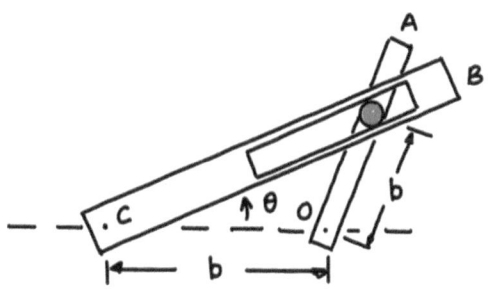

Chapter 8 Notes.

9 Apparent Kinematics

An Introduction. In the previous chapter we developed a kinematic translational velocity relationship for two points on the same rigid body.

$$v_2 = v_1 + \boldsymbol{\omega} \times \boldsymbol{p}_{2/1}$$

Here, v_1 is the translational velocity of point 1, v_2 is the translational velocity of point 2, $\boldsymbol{\omega}$ is the angular velocity of the rigid body, and $\boldsymbol{p}_{2/1}$ is the position vector from point 1 to point 2. A key fact in the development of the above equation is that the distance between points 1 and 2 on the same rigid body is constant.

Here we revisit and extend this development to consider the motion of a point relative to the motion of a rigid body. The new point can be inside or outside of the rigid body; there is no restriction. The process leads to something called an *illusory* or *apparent* or *relative* velocity.

A Simple Motivating Example. To motivate the current study, consider the general planar motion of a satellite. Suppose an antenna is being extended radially outward. Let's develop an expression for the inertial velocity of the antenna tip using the motion of the satellite and the radial extension of the antenna.

The inertial position of the antenna tip, labeled $*$, can be written as the sum of two vectors, $\boldsymbol{p}_* = \boldsymbol{p}_c + \boldsymbol{p}_{*/c}$. The inertial velocity of the tip is simply the inertial time derivative of this expression, $\boldsymbol{v}_* = \boldsymbol{v}_c + \boldsymbol{v}_{*/c}$.

For this simple example, suppose $\boldsymbol{p}_{*/c}$ extends along a body-fixed $\hat{\boldsymbol{b}}_1$ direction, $\boldsymbol{p}_{*/c} = r\hat{\boldsymbol{b}}_1$. The time derivative of $\boldsymbol{p}_{*/c}$ yields $\boldsymbol{v}_{*/c}$, which is computed using the product rule.

$$\boldsymbol{v}_{*/c} = \dot{r}\hat{\boldsymbol{b}}_1 + r\,d/dt(\hat{\boldsymbol{b}}_1) = \dot{r}\hat{\boldsymbol{b}}_1 + r\dot{\theta}\hat{\boldsymbol{b}}_2 = \dot{r}\hat{\boldsymbol{b}}_1 + r\dot{\theta}(\hat{\boldsymbol{b}}_3 \times \hat{\boldsymbol{b}}_1) = \dot{r}\hat{\boldsymbol{b}}_1 + \dot{\theta}\hat{\boldsymbol{b}}_3 \times r\hat{\boldsymbol{b}}_1$$

$$= \dot{r}\hat{\boldsymbol{b}}_1 + \boldsymbol{\omega} \times \boldsymbol{p}_{*/c} \tag{165.1}$$

This may be used in the $\boldsymbol{v}_*$ expression.

$$\boldsymbol{v}_* = \boldsymbol{v}_c + \dot{r}\hat{\boldsymbol{b}}_1 + \boldsymbol{\omega} \times \boldsymbol{p}_{*/c} \tag{165.2}$$

This result deserves a couple of comments:

1. The vector $\boldsymbol{\omega}$ is the angular velocity vector of the rigid body because $\hat{\boldsymbol{b}}_1$ is part of a body-fixed reference frame.
2. The term $\dot{r}\hat{\boldsymbol{b}}_1$ is called $\boldsymbol{v}_{\text{apparent}}$ and is abbreviated as $\boldsymbol{v}_{\text{app}}$. This is the velocity vector of $*$ from the prespective of a body-fixed observer.

$$\boldsymbol{v}_{\text{app}} = \dot{r}\hat{\boldsymbol{b}}_1 \tag{165.3}$$

A body-fixed observer, i.e, one riding *in* and *with* the satellite, would only see the radial distance r growing or shrinking.

Thus, $\boldsymbol{v}_* = \boldsymbol{v}_c + \boldsymbol{v}_{\text{app}} + \boldsymbol{\omega} \times \boldsymbol{p}_{*/c}$ is an expression for the inertial velocity of the antenna tip using the motion of the satellite and the radial extension of the antenna.

⋆ **48 Boom and Slosh Diagrams.**

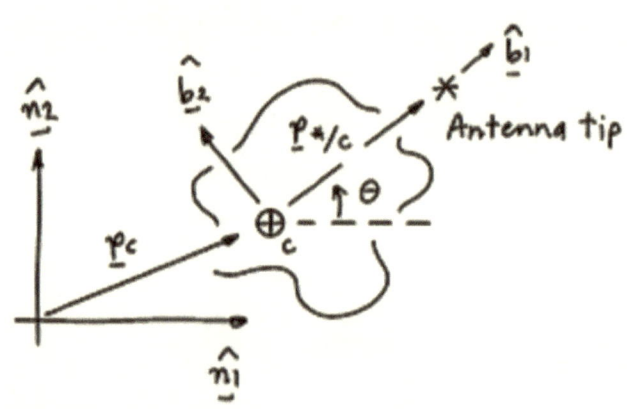

Extending boom tip diagram

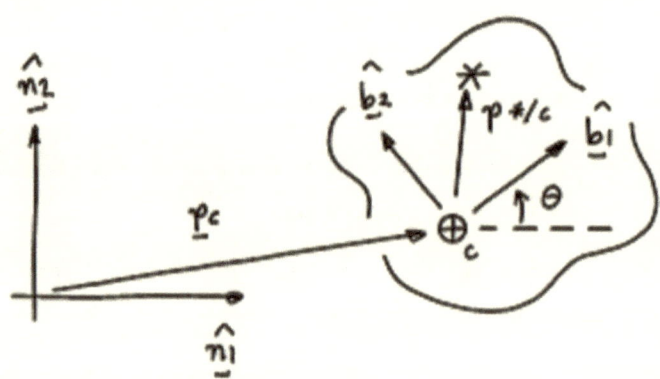

Fluid slosh diagram

Another Example of v_{app}. Let's go further. We now consider the case that a point mass can be anywhere in the plane. A motivating example is the fluid slosh motion of a fuel contained in a satellite.

Suppose the fluid is modeled as a point mass in the rigid body. The inertial position of the fluid point mass representation, labeled $*$, can be written as the sum of two vectors, $\boldsymbol{p}_* = \boldsymbol{p}_c + \boldsymbol{p}_{*/c}$. The inertial velocity of the tip is simply the inertial time derivative of this expression, $\boldsymbol{v}_* = \boldsymbol{v}_c + \boldsymbol{v}_{*/c}$.

The relative position vector $\boldsymbol{p}_{*/c}$ can be written (or expressed or coordinatized) in the body-fixed $\boldsymbol{b}^+$ frame.

$$\boldsymbol{p}_{*/c} = r\hat{\boldsymbol{b}}_1 + s\hat{\boldsymbol{b}}_2 \tag{167.1}$$

The time derivative of this vector yields $\boldsymbol{v}_{*/c}$, and the computation is similar to what was done before.

$$\boldsymbol{v}_{*/c} = \dot{r}\hat{\boldsymbol{b}}_1 + r\,\mathrm{d}/\mathrm{dt}(\hat{\boldsymbol{b}}_1) + \dot{s}\hat{\boldsymbol{b}}_2 + s\,\mathrm{d}/\mathrm{dt}(\hat{\boldsymbol{b}}_2) = \dot{r}\hat{\boldsymbol{b}}_1 + \dot{s}\hat{\boldsymbol{b}}_2 + \boldsymbol{\omega} \times \boldsymbol{p}_{*/c} \tag{167.2}$$

Here we used the truth that $\hat{\boldsymbol{b}}_1$ and $\hat{\boldsymbol{b}}_2$ are parts of a body-fixed reference frame, thus their change with respect to time is related to the rigid body angular velocity vector $\boldsymbol{\omega}$.

The terms $\dot{r}\hat{\boldsymbol{b}}_1 + \dot{s}\hat{\boldsymbol{b}}_2$ compose $\boldsymbol{v}_{\mathrm{apparent}}$, which is the velocity vector of $*$ from the prespective of a body-fixed observer. (That is, one that rides in and with the rigid body.)

$$\boldsymbol{v}_{\mathrm{app}} = \dot{r}\hat{\boldsymbol{b}}_1 + \dot{s}\hat{\boldsymbol{b}}_2 \tag{167.3}$$

Thus, $\boldsymbol{v}_* = \boldsymbol{v}_c + \boldsymbol{v}_{\mathrm{app}} + \boldsymbol{\omega} \times \boldsymbol{p}_{*/c}$ is an expression for the inertial velocity of the fluid slosh using the motion of the satellite and the splish-splash of the fluid.

A Fired Missile and Pull-up Maneuver. As a final example, consider the motion of a fired air-to-air missile from a jet aircraft. We aim to compute the velocity of the missile from the perspective of the pilot. The diagram shows that the inertial position of the point mass missile, labeled $*$, can be written as the sum of two vectors, $\boldsymbol{p}_* = \boldsymbol{p}_c + \boldsymbol{p}_{*/c}$. The inertial velocity of the missile is simply the inertial time derivative of the position vector, $\boldsymbol{v}_* = \boldsymbol{v}_c + \boldsymbol{v}_{*/c}$, which can be rewritten using the previous results.

$$\boldsymbol{v}_* = \boldsymbol{v}_c + \boldsymbol{v}_{\text{app}} + \boldsymbol{\omega} \times \boldsymbol{p}_{*/c} \tag{168.1}$$

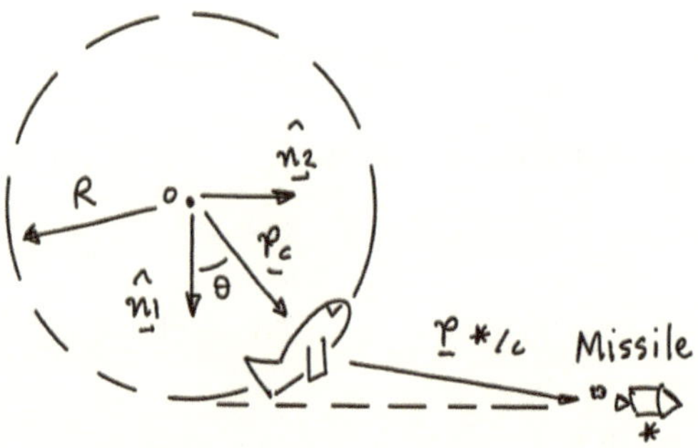

Consider the following data: (1) Let the air-to-air missile travel along a straight horizontal direction, $\boldsymbol{v}_* = \dot{s}\,\hat{\boldsymbol{n}}_2$; (2) Let the jet undergo a pull-up maneuver, $\boldsymbol{v}_c = v\hat{\boldsymbol{b}}_2$; (3) Note that the angular velocity vector of the jet in the pull-up maneuver is related to the jet speed and loop radius, $\boldsymbol{\omega} = (v/R)\,\hat{\boldsymbol{n}}_3$ (the rotation rate equals the revolution rate).

Equation (168.1) can be rearranged to isolate $\boldsymbol{v}_{\text{app}}$.

$$\boldsymbol{v}_{\text{app}} = \boldsymbol{v}_* - \boldsymbol{\omega} \times (\boldsymbol{p}_* - \boldsymbol{p}_c) = \frac{sv}{R}\hat{\boldsymbol{n}}_1 + (\dot{s} - v)\hat{\boldsymbol{n}}_2 \tag{168.2}$$

$$= \left(\frac{sv}{R}\cos\theta - (\dot{s} - v)\sin\theta\right)\hat{\boldsymbol{b}}_1 + \left((\dot{s} - v)\cos\theta - \frac{sv}{R}\sin\theta\right)\hat{\boldsymbol{b}}_2 \tag{168.3}$$

Equation (168.2) is the velocity of the missile from the perspective of the pilot expressed or coordinatized in the inertial frame whereas (168.3) is the same vector expressed or coordinatized in the body-fixed frame.

Visualizing the Missile/Pull-up Example. As an illustration of the missile & pull-up example, consider the special case that missile speed and jet speed are equal, $\dot{s} = v$. One can show that $\boldsymbol{v}_{\text{app}}$ simplifies in this case.

$$\boldsymbol{v}_{\text{app}} = v\theta\cos\theta\hat{\boldsymbol{b}}_1 - v\theta\sin\theta\hat{\boldsymbol{b}}_2 = v\theta\hat{\boldsymbol{n}}_1 \tag{169.1}$$

We can define $\boldsymbol{p}_{\text{app}}$ as the integral (with respect to θ) of these vector components. Also, let $v = 1$ for simplicity.

$$\boldsymbol{p}_{\text{app}} = (\cos\theta + \theta\sin\theta - 1)\hat{\boldsymbol{b}}_1 + (\theta\cos\theta - \sin\theta)\hat{\boldsymbol{b}}_2 \tag{169.2}$$

This is the position vector of the missile from the perspective of a pilot that rides in, and rotates with, the jet.

Interestingly, the $\boldsymbol{p}_{\text{app}}$ vector is the sum of two simple curves that you may recall from your high school geometry class.

$$\boldsymbol{p}_{\text{app}} = \Big((\cos\theta - 1)\hat{\boldsymbol{b}}_1 - \sin\theta\hat{\boldsymbol{b}}_2\Big) + \Big(\theta\sin\theta\hat{\boldsymbol{b}}_1 + \theta\cos\theta\hat{\boldsymbol{b}}_2\Big)$$

$$= (\text{Circle of radius 1 with offset center at } -1)$$

$$+ (\text{Spiral of Archimedes})$$

The top plot on the following page shows the two simple curves. The bottom plot shows a graph of the $\boldsymbol{p}_{\text{app}}$ components in the body-fixed frame. Again, this is the position vector of the missile from the perspective of a pilot that rides in, and rotates with, the jet. Recall that $\hat{\boldsymbol{b}}_1$ is a body-fixed unit vector that points through the floor of the jet, and $\hat{\boldsymbol{b}}_2$ is a body-fixed unit vector that points out the nose of the jet.

The View from the Pilot Seat. These diagrams show the position of the fired missile relative to the jet.

$$\boldsymbol{p}_{\mathrm{app}} = (\cos\theta + \theta\sin\theta - 1)\hat{\boldsymbol{b}}_1 + (\theta\cos\theta - \sin\theta)\hat{\boldsymbol{b}}_2$$

Position (1) has the jet in level flight at the bottom of the loop; position (2) has the jet at $\theta = 90$ deg with its nose up; position (3) has the jet upside down at the top of the loop; position (4) has the jet at $\theta = 270$ deg with its nose down; position (5) has the jet in level flight at the bottom of the loop again; position (6) has the jet at $\theta = 90$ deg with its nose up again; and position (7) has the jet upside down at the top of the loop again.

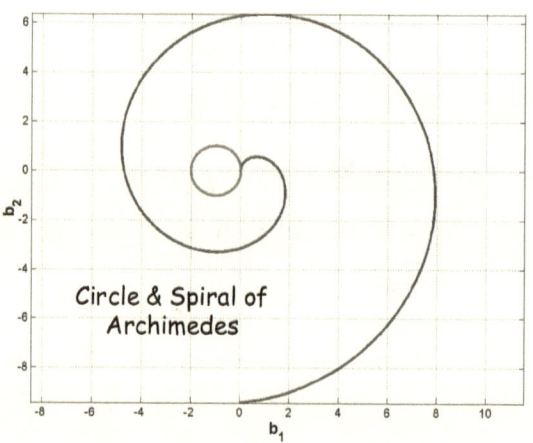

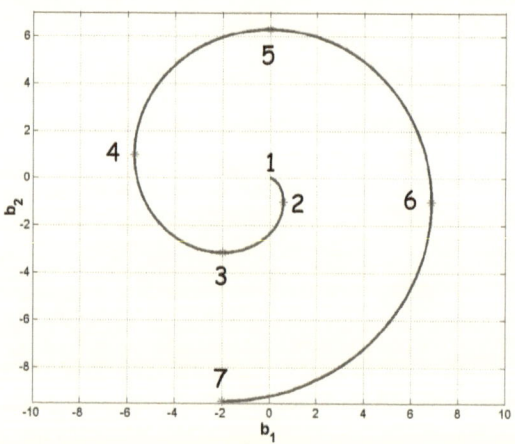

Yo-Yo Despin Kinematics. Satellites are sometimes spun about their axis of symmetry to provide pointing stability. When spin is no longer needed, a so-called yo-yo device can be used to reduce the amount of spin.

The yo-yo device consists of a weight attached to cords wrapped around the satellite in a plane normal to the spin axis. When released, the weights (and cords) unwind in the same direction as the satellite spin. The unwinding angle ϕ grows as time passes, and the length of the cord is related to the unwinding angle, $L = R\phi$. The diagrams below are meant to show this situation. Also shown are portions of three reference frames. Frame $\boldsymbol{n}^+$ is inertial; frame $\boldsymbol{b}^+$ is fixed to the spinning satellite; and frame $\boldsymbol{e}^+$ is such that the unit direction $\hat{e}_1$ always points to the instantaneous connection point of the cable and satellite.

The position vector of the mass relative to the mass center of the satellite is $\boldsymbol{p}_{\text{mass}/c} = R\hat{e}_1 - L\hat{e}_2 = R\hat{e}_1 - R\phi\hat{e}_2$. Understanding a yo-yo despin maneuver requires the velocity vector of the mass relative to the mass center of the satellite as the mass unwinds. This is constructed from the time derivative of $\boldsymbol{p}_{\text{mass}/c}$.

$$\boldsymbol{v}_{\text{mass}/c} = R \, d/dt(\hat{e}_1) - R\dot{\phi}\hat{e}_2 - R\phi \, d/dt(\hat{e}_2)$$

$$= R(\omega + \dot{\phi})\hat{e}_2 - R\dot{\phi}\hat{e}_2 + R\phi(\omega + \dot{\phi})\hat{e}_1 = R\phi(\omega + \dot{\phi})\hat{e}_1 + R\omega\hat{e}_2$$

In developing $\boldsymbol{v}_{\text{mass}/c}$, we've used the fact that the angular velocity of the $\boldsymbol{e}^+$ frame is the combination of the angular velocity of the satellite (or $\boldsymbol{b}^+$ frame) and the angular velocity of the $\boldsymbol{e}^+$ frame relative to the satellite, $\omega_{e/n} = \omega_{e/b} + \omega_{b/n}$.

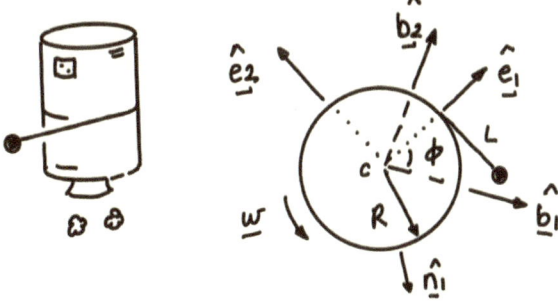

Chapter 9 Notes.

10 Rigid Body Kinetics

An Introduction. Koufax and Drysdale. Simon and Garfunkel. Calvin and Hobbes. Newton and Euler.

Newton gave us $\boldsymbol{f} = m\boldsymbol{a}$. But Euler clarified Newton's meaning and developed laws of mechanics that reach beyond what Newton had in mind. Most noteworthy is Euler's law for rigid body rotational motion. Here, to be symmetric with Newton, we playfully call this Euler's second law, or E2L.

Newton and Euler. Newton's second law of motion was developed for point mass models; Euler clarified the result to show that $\boldsymbol{f} = m\boldsymbol{a}$ holds for the translational motion of rigid bodies, too. The caveat is that when applied in this way, it is the translational motion of the mass center that is described.

$$\boldsymbol{f} = m\boldsymbol{a}_c \qquad (175.1)$$

The vector $\boldsymbol{f}$ is the sum of the external forces acting on the rigid body, m is the total mass, and $\boldsymbol{a}_c$ is the inertial acceleration vector of the rigid body mass center. Equation (175.1) states that the mass center of a rigid body obeys Newton's equation of motion for a point mass. Consequently, this result is occasionally called *the super particle theorem.*

Euler also developed a principle for the rotational motion of a rigid body.

$$\boldsymbol{\ell}_o = \dot{\boldsymbol{h}}_o \qquad (175.2)$$

Here, o is a fixed point, $\boldsymbol{h}_o$ is the rigid body angular momentum vector about o, and $\boldsymbol{\ell}_o$ is the vector sum of external moments (composed of pure moments and those caused by external forces) about o. Equation (175.2) is truly an independent law of mechanics entirely separate from the statement $\boldsymbol{f} = m\dot{\boldsymbol{v}}_c$.

A First Look at Angular Momentum. Euler's law for rotational motion involves the angular momentum vector. The angular momentum vector of a rigid body about a fixed point o is computed by integrating the collection of little momentums over the entire rigid body.

$$\boldsymbol{h}_o \equiv \int \boldsymbol{r} \times \dot{\boldsymbol{r}}\,dm = \int \boldsymbol{r} \times \boldsymbol{v}\,dm \qquad (176.1)$$

A useful expression is the translation theorem for the angular momentum vector. This relates the angular momentum vector about an arbitrary fixed point o to the angular momentum vector about the mass center and motion of the mass center.

$$\boldsymbol{h}_o = \boldsymbol{p}_c \times m\boldsymbol{v}_c + \boldsymbol{h}_c \qquad \text{where} \qquad \boldsymbol{h}_c \equiv \int \boldsymbol{\rho} \times \dot{\boldsymbol{\rho}}\,dm \qquad (176.2)$$

Here, $\boldsymbol{\rho}$ is the relative position vector of a point in the rigid body measured from the mass center and $\dot{\boldsymbol{\rho}}$ is the relative velocity vector between these two points. Essential use of $\int \boldsymbol{\rho}\,dm = 0$ is used in this development: this expression is true because $\boldsymbol{\rho}$ is measured from the mass center.

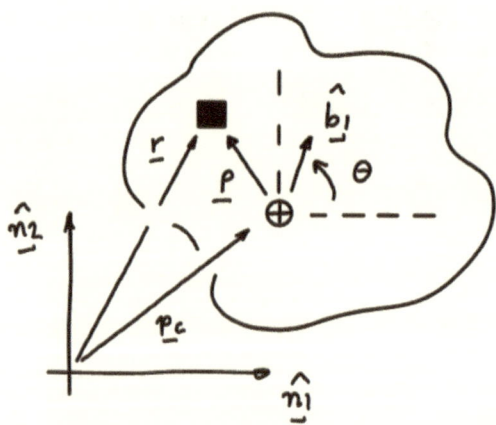

⋆ 49 Showing the Translation Theorem for Angular Momentum.

Begin with the definition of angular momentum.

$$\underline{h}_0 = \int \underline{r} \times \underline{v} \; dm$$

But $\underline{r} = \underline{p}_c + \underline{\rho}$.

And because these are two points on the same rigid body,

$$\underline{v} = \underline{v}_c + d/dt\,(\underline{\rho}) = \underline{v}_c + \underline{\omega} \times \underline{\rho}$$

so $\underline{h}_0 = \int (\underline{p}_c + \underline{\rho}) \times (\underline{v}_c + d/dt\,(\underline{\rho}))\, dm$

$$= \int (\underline{p}_c \times \underline{v}_c)\, dm + \int \underline{\rho} \times \underline{v}_c \; dm$$

$$+ \int \underline{p}_c \times d/dt\,(\underline{\rho})\, dm + \int \underline{\rho} \times d/dt\,(\underline{\rho})\, dm$$

$$\underline{h}_0 = \underline{p}_c \times m\,\underline{v}_c + \int \underline{\rho}\, dm \times \underline{v}_c$$

$$+ \underline{p}_c \times d/dt\left(\int \underline{\rho}\, dm\right) + \int \underline{\rho} \times d/dt\,(\underline{\rho})\, dm$$

But $\int \underline{\rho}\, dm = 0$ so

$$\underline{h}_0 = \underline{p}_c \times m\,\underline{v}_c + \int \underline{\rho} \times d/dt\,(\underline{\rho})\, dm$$

A Collection of Newton and Euler Laws. The translation theorem for the angular momentum vector relates the angular momentum vector about a fixed point to the angular momentum vector about the mass center and the translational motion of the mass center.

$$\boldsymbol{h}_o = \boldsymbol{p}_c \times m\boldsymbol{v}_c + \boldsymbol{h}_c \qquad \text{where} \qquad \boldsymbol{h}_c \equiv \int \boldsymbol{\rho} \times \dot{\boldsymbol{\rho}}\, dm \qquad (178.1)$$

We won't show it here, but this expression, together with the elementary concept that an externally applied force produces a couple via $\boldsymbol{r} \times \boldsymbol{f}$, gives another form of Euler's principle that is as simple as the original: $\boldsymbol{\ell}_c = \dot{\boldsymbol{h}}_c$.

Written together, the Newton-Euler laws of mechanics for rigid bodies are the following.

Newton's law for the mass center: $\qquad$ N2L$_c$ $\quad \boldsymbol{f} = m\boldsymbol{a}_c$

Euler's law about the mass center: $\qquad$ E2L$_c$ $\quad \boldsymbol{\ell}_c = \dot{\boldsymbol{h}}_c$

Euler's law about a fixed point: $\qquad$ E2L$_o$ $\quad \boldsymbol{\ell}_o = \dot{\boldsymbol{h}}_o$

The first and second equations are useful for general rigid body motion. The third equation is useful for rotational motion of a rigid body about a fixed point.

Euler deserves much credit.

Introducing Inertia. The motion of a point mass model depends on the mass of the system. The planar motion of a rigid body depends on more. It depends on how the mass is distributed. Let's see.

The angular momentum vector of a rigid body about its mass center was introduced earlier.

$$h_c \equiv \int \rho \times \dot\rho \, dm \qquad (179.1)$$

The kinematic expressions that relate two points on the same rigid body had a role in coming to this result.

$$r = p_c + \rho \quad ; \quad v = v_c + \dot\rho = v_c + \omega \times \rho \qquad (179.2)$$

Here, ω is the rigid body angular velocity vector. Importantly, $\dot\rho = \omega \times \rho$, so $h_c = \int \rho \times (\omega \times \rho) \, dm$. The following page will show that this simplifies for rigid bodies undergoing planar motions.

$$h_c = I_c \omega \, \hat{b}_3 \qquad (179.3)$$

Here, I_c is the rigid body inertia (or rigid body mass moment of inertia) in the plane about the mass center. For rigid bodies, I_c is constant, just like the mass m is constant.

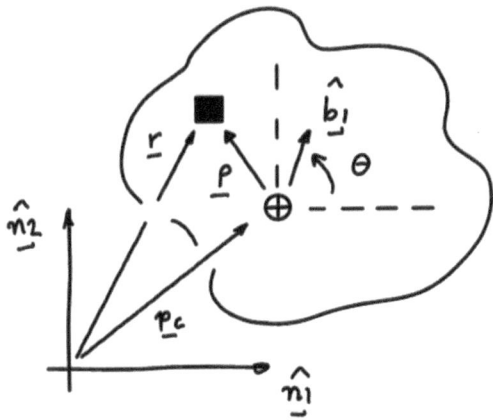

⋆ 50 Rigid Body Inertia in the Plane.

Expressing the vector $\underline{\rho}$ in a body-fixed plane,

$$\underline{\rho} = \rho_x \, \hat{\underline{b}}_1 + \rho_y \, \hat{\underline{b}}_2$$

The angular velocity vector is

$$\underline{\omega} = \omega \, \hat{\underline{b}}_3$$

so $\underline{h}_c = \int \underline{\rho} \times (\underline{\omega} \times \underline{\rho}) \, dm$

$$= \int (\rho_x \, \hat{\underline{b}}_1 + \rho_y \, \hat{\underline{b}}_2) \times (-\omega \rho_y \, \hat{\underline{b}}_1 + \omega \rho_x \, \hat{\underline{b}}_2) \, dm$$

$$= \int (\rho_x^2 + \rho_y^2) \, \omega \, dm \; \hat{\underline{b}}_3$$

$$= \left\{ \int (\rho_x^2 + \rho_y^2) \, dm \right\} \, \omega \, \hat{\underline{b}}_3$$

Define $\quad I_c \equiv \int (\rho_x^2 + \rho_y^2) \, dm$

$$\underline{h}_c = I_c \, \omega \, \hat{\underline{b}}_3$$

I_c is the inertia in the plane about the mass center.

⋆ **51 Planar Inertia of Common Shapes.**

Slender Rod of length L

$$I_c = \frac{1}{12} m L^2$$

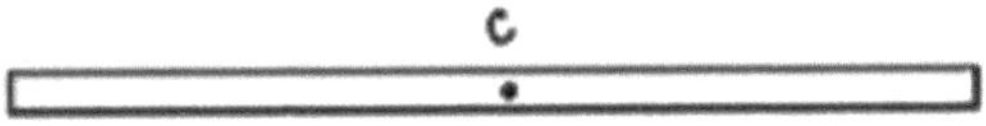

c

Flat Plate

$$I_c = \frac{1}{12} m (a^2 + b^2)$$

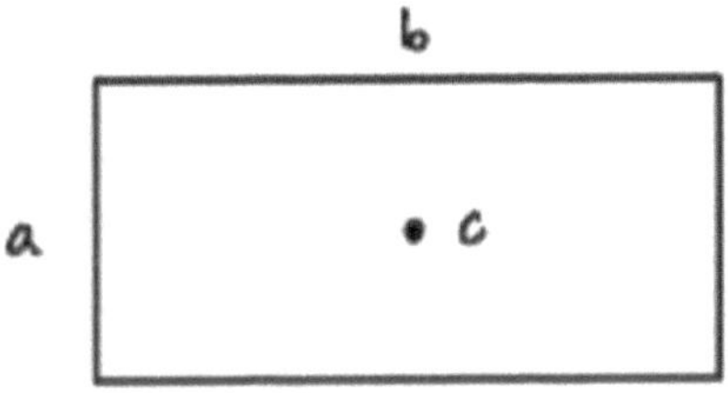

b

a

• c

Solid Circular Disk

$$I_c = \frac{1}{2} m R^2$$

R

c

Hollow Circular Ring

$$I_c = m R^2$$

R

c

Solid Sphere

$$I_c = \frac{2}{5} m R^2$$

R

c

The Useful Collection of Newton and Euler Laws. Equations can be written differently now that rigid body inertia in the plane about the mass center has been defined.

$$\boldsymbol{h}_c = I_c \omega \, \hat{\boldsymbol{b}}_3 \quad \rightarrow \quad \dot{\boldsymbol{h}}_c = I_c \alpha \, \hat{\boldsymbol{b}}_3 \qquad (182.1)$$

Similarly, if point o is a fixed point *on the rigid body*, then one can show that rigid body inertia in the plane about the fixed point o comes into play.[8]

$$\boldsymbol{h}_o = I_o \omega \, \hat{\boldsymbol{b}}_3 \quad \rightarrow \quad \dot{\boldsymbol{h}}_o = I_o \alpha \, \hat{\boldsymbol{b}}_3 \qquad (182.2)$$

Consequently, the Newton-Euler laws of mechanics for the planar motion of rigid bodies are as follows.

Newton's law for the mass center:	N2L$_c$ $\boldsymbol{f} = m\boldsymbol{a}_c$
Euler's law about the mass center:	E2L$_c$ $\boldsymbol{\ell}_c = I_c \alpha \, \hat{\boldsymbol{b}}_3$
Euler's law about a fixed point on the body:	E2L$_o$ $\boldsymbol{\ell}_o = I_o \alpha \, \hat{\boldsymbol{b}}_3$

The first and second equations are useful for general rigid body motion. The third equation is useful for rotational motion of a rigid body about a fixed point on the rigid body.

---

[8] The real key to this result is that the revolution rate equals the rotation rate, which is true for rotation about a fixed point on a body.

A Rigid Body Routine. The Newton-Euler laws of mechanics produce equations that govern the motion of rigid bodies. A procedure for generating the equations can be established.

1. Identify the system.

2. Draw a free body diagram (FBD).

3. Establish relevant reference frames.

4. Write a vector representation of the applied forces and moments.

5. Perform the vector kinematics: determine an expression for the translational acceleration vector of the mass center; determine an expression for the rotational angular acceleration.

6. Use $N2L_c$ and $E2L_c$ to write the governing equations of motion. If the body is rotating about a fixed point on the body, one can use $E2L_o$ instead.

⋆ 52 A Swinging Rod.

A rigid bar that may swing about the fixed point O is released from rest. The bar has length L and mass m. Determine the governing equations of motion. Compute the reactions at O and the angular acceleration vector.

Steps 1, 2, 3, 4.   System, FBD, Reference frames and vector representation of the forces.

$$\underline{f} = A_x \,\hat{b_1} - A_y \,\hat{b_2} + W \,\hat{b_2}$$

$$\underline{l}_c = A_y \frac{L}{2} \,\hat{b_3}$$

Step 5.
Kinematics

$$\underline{p}_c = L/2 \,\hat{b_1} \quad ; \quad \underline{v}_c = L/2 \,\dot{\theta} \,\hat{b_2}$$

$$\underline{a}_c = -\frac{L}{2} \dot{\theta}^2 \,\hat{b_1} + \frac{L}{2} \ddot{\theta} \,\hat{b_2}$$

Since the bar begins from rest $\underline{a}_c = \frac{L}{2} \alpha \,\hat{b_2}$

Also $\underline{\alpha} = \alpha \,\hat{b_3}$

Step 6. $N2L_c$ and $E2L_c$

$$Ax = 0 \quad ; \quad mg - Ay = m \, ^{L/2} \alpha$$

$$Ay \, ^{L/2} \, \hat{b_3} = I_c \, \alpha \, \hat{b_3}$$

But $I_c = \frac{1}{12} m L^2$ so $Ay = \frac{1}{6} mL \, \alpha$

$N2L_c$ and $E2L_c$ provide 3 equations for 3 unknowns.

(Ax, Ay, α)

$$Ax = 0 \quad ; \quad \alpha = \frac{3}{2} \frac{g}{L} \quad ; \quad Ay = \frac{W}{4}$$

Alternatively, one could investigate $E2L_0$

$$\underline{l_0} = I_0 \, \alpha \, \hat{b_3}$$

$$\underline{l_0} = mg \, L/2 \, \hat{b_3}$$

$$I_0 = I_c + m \left(^{L/2} \right)^2 = \frac{1}{3} m L^2$$

so $mg \frac{L}{2} = \frac{1}{3} m L^2 \alpha$ or $\alpha = \frac{3}{2} \frac{g}{L}$

☐ Same answer for α as before

☐ α obtained more directly

☐ still need $\underline{f} = m \underline{a_c}$ to determine Ax and Ay

186

Parallel Axis Theorem and Radius of Gyration. The rigid body inertia in the plane about the mass center for some common geometric shapes was presented earlier. In our notation, this is I_c. The parallel axis theorem is an expression that relates the inertia about an arbitrary fixed point o in the plane to I_c and the distance between o and c.

$$I_o = I_c + md^2 \tag{186.1}$$

This is straightforward to derive beginning with a definition of inertia about a point and replacing distances with a distance to the mass center. The computations are similar to what we've done before.

$$
\begin{aligned}
I_o &= \int (\delta_x^2 + \delta_y^2)\, dm \quad \text{where} \quad \boldsymbol{\delta} = \delta_x \hat{\boldsymbol{b}}_1 + \delta_y \hat{\boldsymbol{b}}_2 = (d_x + \rho_x)\hat{\boldsymbol{b}}_1 + (d_y + \rho_y)\hat{\boldsymbol{b}}_2 \\
&= \int \left(d_x^2 + \rho_x^2 + 2d_x\rho_x + d_y^2 + \rho_y^2 + 2d_y\rho_y \right) dm \\
&= I_c + m\left(d_x^2 + d_y^2 \right) = I_c + md^2
\end{aligned}
\tag{186.2}
$$

The parallel axis theorem was used in the previous example when E2L$_o$ was investigated.

Another useful inertia expression involves the radius of gyration, k. This is a special value such that the rigid body inertia in the plane satisfies $I = mk^2$. A radius of gyration is some times given when a rigid body is composed of different materials.

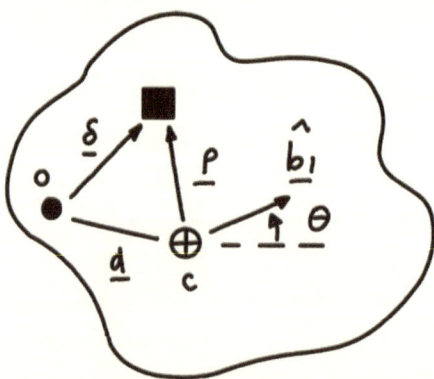

⋆ 53 General Planar Motion of a Rod.

A rigid bar that may slide without friction on a horizontal table is at rest. The bar has length L and mass m. A force F acts at one end. Determine the governing equations of motion. Compute the acceleration of end A.

{ Gravity acts into the page }

A   C $\downarrow F$

Steps 1, 2, 3, 4.   System, FBD, Reference frames and vector representation of the forces.

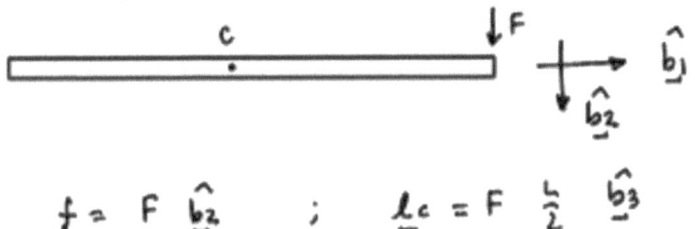

$$\underline{f} = F\ \hat{b_2} \qquad ; \qquad \underline{l}_c = F\ \tfrac{L}{2}\ \hat{b_3}$$

step 5. kinematics

$$\underline{a}_c = a_x\ \hat{b_1} + a_y\ \hat{b_2} \qquad ; \qquad \underline{\alpha} = \alpha\ \hat{b_3}$$

step 6. N2L$_G$ and E2L$_c$

$$F = m\ a_y \qquad ; \qquad 0 = m\ a_x$$

$$F\ \frac{L}{2} = I_c\ \alpha = \frac{1}{12}\ m L^2\ \alpha$$

188

Solving

$$a_x = 0 \quad ; \quad a_y = F/m \quad ; \quad \alpha = \frac{Fb}{ML}$$

Now, points A and C are two points on the same rigid body. So,

$$\underline{a}_A = \underline{a}_C + \underline{\alpha} \times \underline{r}_{A/C} + \underline{\omega} \times (\underline{\omega} \times \underline{r}_{A/C})$$

But $\underline{\omega} = 0$ because bar begins from rest.

$$\underline{a}_A = \frac{F}{m}\,\hat{b_2} + \frac{Fb}{ML}\,\hat{b_3} \times (-\tfrac{L}{2}\,\hat{b_1})$$

$$= \frac{F}{m}\,\hat{b_2} - \frac{3F}{m}\,\hat{b_2}$$

$$= -2\frac{F}{m}\,\hat{b_2}$$

★ 54 Two Rigid Bodies.

A disk and bar are shown. A moment M is applied to the disk. Friction acts between the disk and incline. Derive the governing equations of motion.

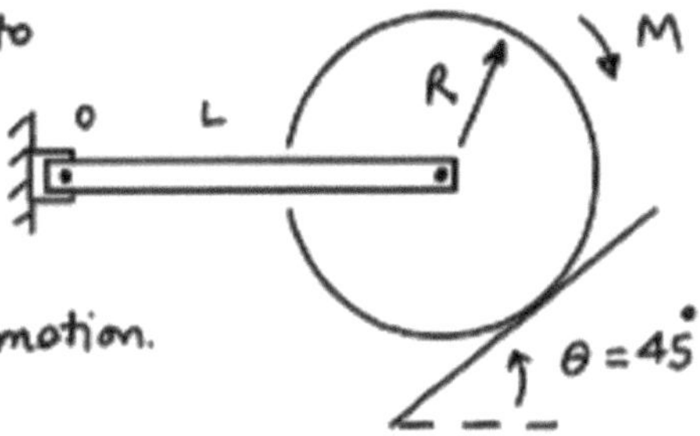

Steps 1, 2, 3, 4

(system, FBD, ref frames, vector rep.)

Disk

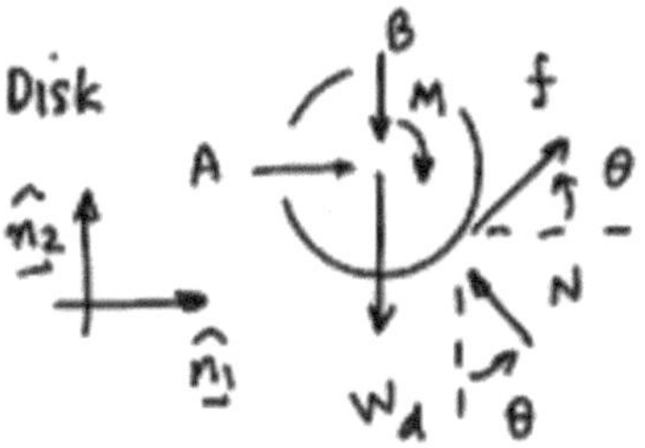

$$\underline{F} = -W_d \,\hat{n_2} + N\cos\theta \,\hat{n_2}$$
$$\qquad - N\sin\theta \,\hat{n_1} + f\cos\theta \,\hat{n_1}$$
$$\qquad + f\sin\theta \,\hat{n_2}$$
$$\qquad + A\,\hat{n_1} - B\,\hat{n_2}$$

$$\underline{\ell_c} = -M\,\hat{n_3} + Rf\,\hat{n_3}$$

Step 5. Kinematics

$$\underline{a_c} = 0 \quad ; \quad \underline{\alpha} = \alpha\,\hat{n_3}$$

Step 6. N2L_c and E2L_c

N2L_c : $-W_d + N\cos\theta + f\sin\theta - B = 0$ ①

$\qquad -N\sin\theta + f\cos\theta + A = 0$ ②

E2L_c : $-M + Rf = \frac{1}{2}mR^2\alpha$ ③

But $f = \mu_k N$ ④

Steps 1, 2, 3, 4

(system, FBD, ref frames, vector rep.)

Bar :

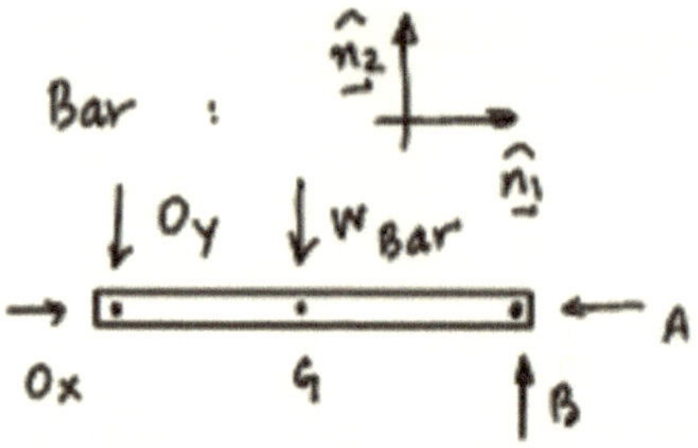

$$\underline{F}_{Bar} = (Ox - A)\,\hat{n}_1$$
$$+ (B - Oy)\,\hat{n}_2$$
$$- W_{Bar}\,\hat{n}_2$$

$$\underline{\ell}^G_{Bar} = (B + Oy)\frac{L}{2}\,\hat{n}_3$$

$$\underline{\ell}^O_{Bar} = (BL - W_{Bar}\frac{L}{2})\,\hat{n}_3$$

step 5. kinematics of bar : $\underline{a}_G = 0$; $\underline{\alpha} = 0$

step 6. N2L$_G$ and E2L$_O$

E2L$_O$: $BL - W_{Bar}\,L/2 = 0$ ⑤

N2L$_G$: $Ox - A = 0$ ⑥

$B - Oy - W_{Bar} = 0$ ⑦

7 linear algebraic Eqns

7 unknowns — N, f, B, A, α, Ox, Oy

Example results :

$$N = \frac{\sqrt{2}}{2}\,(W_{Bar} + 2W_d) \div (1 + \mu_k)$$

$$A = \frac{1}{2}\,\frac{(1 - \mu_k)\,(W_{Bar} + 2W_d)}{(1 + \mu_k)}$$

$$\alpha = \sqrt{2}\, g \left\{ - \sqrt{2}\, M \left(1 + \mu_k \right) + \mu_k\, R \left(W_{Bar} + 2 W_d \right) \right\}$$

$$W_d\ R^2\ \left(1 + \mu_k \right)$$

Some numerical parameters & answers.

mass Disk = 10 kg;
g = 10 m/sec^2;
R = .2 m;
muk = 0.4;
Moment = 10 Nm;
mass Bar = 1 kg;
theta = pi/4 rad;

N = 106.0660 Newtons
f = 42.4264 Newtons
B = 5.0000 Newtons
A = 45.0000 Newtons
alpha = -7.5736 rad/sec^2
Ox = 45.0000 Newtons
Oy = -5.0000 Newtons

Will the Wheel Slide?   Sometimes a wheel is in contact with a surface and acted on by forces, and it is not evident what kind of motion will happen: will the wheel roll without slip or will it slide? Below is a method to answer such questions.

1. Solve the equations assuming the wheel rolls without slip. This means there is a relationship between the acceleration of the mass center and the angular acceleration of the wheel. Compute the (necessary) friction force, f_{necess}, that is consistent with this assumption. Confirm that $f_{necess} > 0$. Compare the (necessary) friction force with the maximum friction force available, $f_{max} = \mu_s N$.

   ○ If $f_{necess} \leq f_{max}$ then the original assumption was correct. The wheel rolls without slip. We are done.

   ○ If $f_{necess} > f_{max}$ then the original assumption was incorrect. Sliding occurs and the acceleration of the mass center and the angular acceleration of the wheel are not related after all. Continue to the next step.

2. Re-solve the problem assuming that sliding takes place. This means $f_{kin} = \mu_k N$ and we must now use this friction model. The acceleration of the mass center and the angular acceleration of the wheel are independent.

   This approach is used in the next example.

⋆ 55 Pulling on a Disk.

A spool is pulled with force F.
The mass of the spool is m and
the inertia about C is I_c.
Compute a_c and $\underline{\alpha}$.

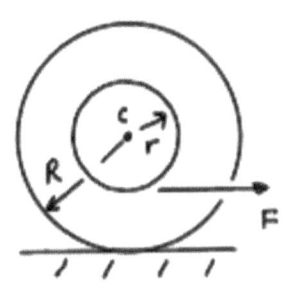

$m = 5 \text{ kg}$

$F = 20 \text{ N}$ $I_c = 5 \cdot (.12)^2$

$\mu_s = 1/4$ $r = .08 \text{ m}$

$\mu_k = 1/5$ $R = 0.16 \text{ m}$

steps 1,2,3,4. (System, FBD, ref frames, vector rep.)

$$\underline{R} = F\,\hat{n}_1 - f\,\hat{n}_1$$
$$\quad + N\,\hat{n}_2 - W\,\hat{n}_2$$

$$\underline{l}_c = (Fr - fR)\,\hat{n}_3$$

step 5. kinematics. Assume roll without slip.

$$\underline{\alpha} = -\alpha\,\hat{n}_3 \quad ; \quad \underline{a}_c = R\,\alpha\,\hat{n}_1 \qquad \{\text{forces } F$$

$$\text{and } N \text{ are}$$

step 6. N2L$_c$ & E2L$_c$

$$\text{independent }\}$$

N2L$_c$: $F - f = m R \alpha$

$$N - W = 0$$

E2L$_c$: $(Fr - fR) = -I_c \alpha$

These can be solved $f = F\left(\dfrac{I_c + mRr}{I_c + mR^2}\right) = 13.6 \text{ N}$

This is the f consistent with roll without slip. Compare to f_{max}.

$$f_{max} = \mu_s N = \tfrac{1}{4}(5)(4.81) = 12.3 \; N$$

Since $f > f_{max}$ the rigid body slides.

so $f = \mu_k N$ & no-slip kinematic constraint is not valid. Thus, $\underline{a}_c$ & $\underline{\alpha}$ are independent.

Let $\underline{a}_c = a \; \hat{n}\hat{j}$ and revisit the equations of motion.

N2L$_c$: $\quad F - f = m a$
$$N - W = 0$$

E$\mathcal{L}$L$_c$: $\quad (Fr - fR) = -I_c \alpha$
$$f = \tfrac{1}{5} N$$

Solving, $a = 2.04 \, \dfrac{m}{s^2}$; $\alpha = -0.42 \, \dfrac{rad}{s^2}$

⋆ 56 The 'Backspinning' Ball.

A bowling ball is released with a forward velocity v_0 and a backward spin w_0. Let μ_k be the kinetic friction value. Compute t_* when the ball rolls without slip. What is the translational speed at time t_*? Parameters are

$m = 5 \text{ kg}$; $R = 0.1 \text{ m}$; $v_0 = 5 \text{ m/s}$
$w_0 = 10 \text{ rad/sec}$; $\mu_k = 0.10$

Steps 1,2,3,4. (System, FBD, ref frames, vector rep.)

$$\underline{F} = -f \,\hat{n}_1$$
$$+ N \,\hat{n}_2 - w \,\hat{n}_2$$

$$\underline{l}_c = -f R \,\hat{n}_3$$

step 5. kinematics

$$\underline{a}_c = -a \,\hat{n}_1$$
$$\underline{\alpha} = -\alpha \,\hat{n}_3$$

step 6. N2L$_c$ and E2L$_c$

N2L$_c$: $-f = -m a$
$N - W = 0$

E2L$_c$: $-f R = -I_c \alpha$

& $f = \mu_k N$

Solving

$a = \mu_k g$

$\alpha = \dfrac{5}{2} \dfrac{\mu_k g}{R}$

Note that a and α are constant. Thus the velocities can be written.

$$\underline{v} = v\, \hat{n}_1 \quad \text{so}$$

$$v = v_0 - at = v_0 - \mu_k g\, t$$

And $\underline{w} = w\, \hat{n}_3$ so

$$w = w_0 - \alpha t = w_0 - \frac{5}{2}\, \frac{\mu_k g}{R}\, t$$

No slip occurs when $v = -Rw$

so,

$$v_0 - \mu_k g\, t_* = -Rw_0 + \frac{5}{2}\, \mu_k g\, t_*$$

$$t_* = \frac{v_0 + Rw_0}{\mu_k g} \cdot \frac{2}{7} = 1.75 \text{ sec.}$$

$$v(t_*) = v_0 - \frac{2}{7}(v_0 + Rw_0)$$

$$= \frac{5}{7}v_0 - \frac{2}{7}Rw_0 = 3.29 \text{ m/s}$$

{ Can also determine when $w = 0$.

$$t_a = \frac{2}{5}\, \frac{Rw_0}{\mu_k g} = 0.41 \text{ sec } \}$$

(Hula Hoops and Bowling Balls). Hula hoops and bowling balls can be set into motion with a simultaneous forward velocity and backspin. Once all of the slipping and sliding is over, bowling balls continue their forward motion to knock down pins whereas hula hoops return to the hand launched them. Why is that?

Both systems are governed by the same Newton/Euler equations of motion, and one can find solutions for the translational acceleration of the mass center and the angular acceleration: $a = \mu_k g$, $\alpha = \mu_k g r / k^2$. Here, μ_k is the kinetic friction coefficient, g is the gravitational acceleration, r is the rigid body radius, and k is the radius of gyration.

The velocity level kinematics can be obtained through integration.

$$v = v_0 - \mu_k g t \quad ; \quad \omega = \omega_0 - \frac{\mu_k g r}{k^2} t \tag{197.1}$$

From here, one can compute the time that the translational velocity may go to zero, and the time that the angular velocity may go to zero.

$$t_{(v=0)} = \frac{v_0}{\mu_k g} \quad ; \quad t_{(\omega=0)} = \frac{\omega_0 k^2}{\mu_k g r} \tag{197.2}$$

These expressions lead to the famously-known *backspin coefficient*.

$$\varpi = \frac{\omega_0 k^2}{v_0 r} \tag{197.3}$$

If $\varpi > 1$ then the circular rigid body returns to the hand that launched it ... think hula hoop (v goes to zero before ω).

If $\varpi < 1$ then the circular rigid body continues forward ... think bowling ball (ω goes to zero before v).

If $\varpi = 1$ then the circular rigid body will neither continue forward or return: it will come to a dead stop (v and ω achieve zero at the same time).

Chapter 10 Problem Set.

1. (M&K 6.25) The testing of a lunar module model is done in a lab using a pendulum support system consisting of parallel cables A and B. The pendulum system allows the model to experience no rotational motion as it swings. The model has a mass of 10 kg with a mass center at G. (a) Use the six step process to develop the governing equations of motion for the lunar module model. You should find three scalar equations from N2L$_c$ and E2L$_c$. A rotating reference frame with a unit vector that is parallel to the cables is useful. (b) Suppose at a particular instant $\dot{\theta} = 2$ rad/sec and $\theta = 60$ deg. Compute the tension in each cable at this instant. Answer: $T_B = 147.9$ N.

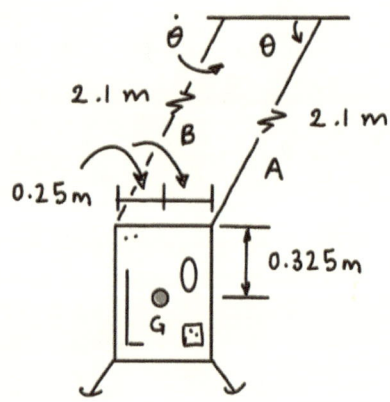

2. (M&K 6.62) A semicircular disk is free to pivot about a fixed bearing at point o. The disk has mass m and radius r. Moreover, the moment of inertia about the fixed point is $I_o = \frac{1}{2}mr^2$, and the distance from o to the mass center is $r_{c/o} = \frac{4r}{3\pi}$. (a) Use E2L$_o$ and N2L$_c$ to develop the governing equations of motion. (b) Use the chain rule to partially solve these equations and find an expression that relates the angular velocity of the disk to the angular displacement of the disk. In doing this, assume that the disk begins from rest when $\theta = 0$. (c) Derive expressions for the normal and tangential forces on the bearing as a function of θ. Answer: $F_n = \left(1 + 64/(9\pi^2)\right) mg \sin \theta$.

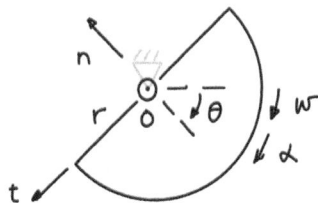

3. (M&K 6.78) A Sandia missile is at an altitude of 400 km where the acceleration due to gravity is 8.69 m/s². The missile has a remaining mass of 300 kg and is pointed 30 deg from the vertical direction. Suddenly, the engine ignites producing a thrust T equal to 4 kN through a nozzle that is tilted by 1 deg from the axis of symmetry. Point c marks the center of mass, and the body has a radius of gyration equal to 1.5 m. (a) Use E2L$_c$ and N2L$_c$ to develop the governing equations of motion. (b) Solve for the angular acceleration of the missile at this instant. (c) Solve for the $\hat{n}_1$ and $\hat{n}_2$ components of the translational acceleration of the mass center c. Answer: $a_c = (6.87\hat{n}_1 + 2.74\hat{n}_2)$ m/s².

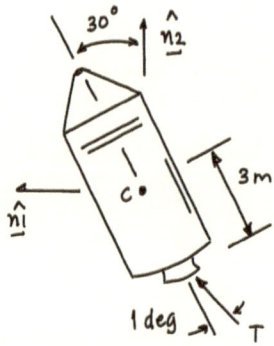

4. For simplicity a lunar rover is modeled as a large rigid hoop with radius R and mass m. The inertia about the mass center is $I_c = mR^2$. NASA engineers are tasked with computing the largest instantaneous moment M that can be applied about the mass center so that the hoop, beginning from rest, rolls over (climbs over) a small rock in a manner that is consistent with roll-without-slip. The small rock is modeled as a stationary sphere of radius r. Gravity acts downward in the sketch. There is friction at the contact point between the hoop and sphere with coefficients μ_s and μ_k, respectively. Note that the forces on the hoop from contact with the horizontal surface will go to zero at the instant that the hoop begins its climb. (a) Use E2L$_c$ and N2L$_c$ to fully and clearly develop the governing equations for the rigid hoop that are valid at the instant the hoop begins its climb. You should find three scalar equations. (b) Solve for an expression that gives the largest moment M that can be applied at the instant the hoop begins its climb. Show that your answer is $M = 2mgR\left[\,(R-r)\mu_s - \sqrt{rR}\,\right]/(R+r)$.

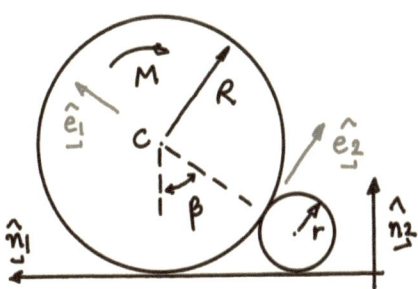

5. A sphere moves along an inclined plane. The inertia of a sphere about its mass center is $I_c = (2/5)mr^2$. Assume the angular acceleration is $\alpha = -\alpha\hat{n}_3$ and the angular velocity is $\omega = -\omega\hat{n}_3$. Friction acts between the incline and sphere. The coefficients of friction are $\mu_s = 0.1$ and $\mu_k = 0.05$. The kinematic relationship between the translational acceleration of the mass center and angular acceleration of the sphere, if the sphere does not slide, is $a_c = r\alpha$. (a) Use E2L$_c$ and N2L$_c$ to fully and clearly develop the governing equations for the rigid sphere. (b) Determine if the sphere slides or not. Show all work.

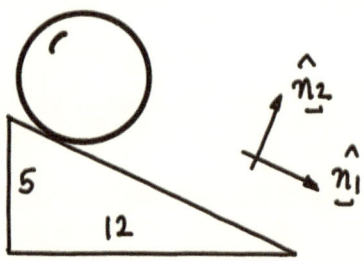

6. (M&K 6.109) Consider the planar motion of a rigid bar. The uniform bar has mass m and length ℓ. The inertia about the mass center is $I_c = (1/12)m\ell^2$. One end of the bar is constrained to move along a horizontal surface. There is no friction between the bar and surface, and a gravity acts downward. Our goal is to develop and analyze the equations of motion.

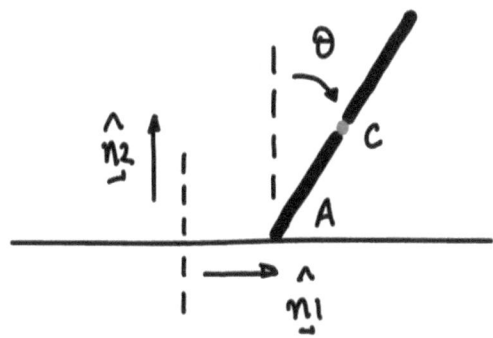

(a) A FBD of the bar shows a vertical normal force from contact with the surface acting at end A, and a vertical force representing the weight acting at the mass center C. Write a vector representation of these forces in the n^+ frame. Also write a vector representation of moments these forces produce about the mass center.

Measuring from a fixed location, the position vector of the mass center may be written using Cartesian coordinates as $p_C = x\hat{n}_1 + y\hat{n}_2$. The translational velocity and acceleration vectors of the mass center are $v_C = \dot{x}\hat{n}_1 + \dot{y}\hat{n}_2$, and $a_C = \ddot{x}\hat{n}_1 + \ddot{y}\hat{n}_2 = a_1\hat{n}_1 + a_2\hat{n}_2$. Measuring from the vertical, the angular position can be measured with the angle θ. The angular velocity and angular acceleration vectors can be taken as $\omega = -\dot{\theta}\hat{n}_3 = -\omega\hat{n}_3$ and $\alpha = -\ddot{\theta}\hat{n}_3 = -\alpha\hat{n}_3$.

(b) Use E2L$_c$ and N2L$_c$ to develop three scalar equations of motion. You should find the following:

$$0 = ma_1 \quad ; \quad N - mg = ma_2 \quad ; \quad \frac{1}{2}N\ell\sin\theta = \frac{1}{12}m\ell^2\alpha$$

(c) Recall end A is constrained to move along the horizontal surface. Let the acceleration vector of end A be taken as $\mathbf{a}_A = -a\hat{n}_1$. Use vector kinematics of two points on a rigid body to show the following:

$$\mathbf{a}_C = (-a + \frac{1}{2}\ddot{\theta}\ell\cos\theta - \frac{1}{2}\dot{\theta}^2\ell\sin\theta)\hat{n}_1 + (-\frac{1}{2}\ddot{\theta}\ell\sin\theta - \frac{1}{2}\dot{\theta}^2\ell\cos\theta)\hat{n}_2$$

(d) Substitute the results from part (c) into the results from part (b) to re-write the equations of motion.

$$0 = m(-a + \frac{1}{2}\ddot{\theta}\ell\cos\theta - \frac{1}{2}\dot{\theta}^2\ell\sin\theta)$$

$$N - mg = m(-\frac{1}{2}\ddot{\theta}\ell\sin\theta - \frac{1}{2}\dot{\theta}^2\ell\cos\theta)$$

$$\frac{1}{2}N\ell\sin\theta = \frac{1}{12}m\ell^2\ddot{\theta}$$

(e) Use the previous results to relate the acceleration of end A to the angular motion of the bar. Do the same for the translational acceleration of the mass center. Answer: $\mathbf{a}_C = (-\frac{1}{2}\ddot{\theta}\ell\sin\theta - \frac{1}{2}\dot{\theta}^2\ell\cos\theta)\hat{n}_2$.

(f) Use the results from part (d) to find a differential equation for θ. Answer: $\ddot{\theta} = (2g/\ell - \dot{\theta}^2\cos\theta)\sin\theta/(\sin^2\theta + 1/3)$.

(g) The solution for the translational motion of the mass center can be computed once the solution for θ is determined. Show that the vertical position of the mass center is $y = (\ell/2)\cos\theta + c_1 t + c_2$, where c_1 and c_2 are constants of integration.

(h) Use Euler's simple numerical integration technique to numerically compute the solution of the differential equation for θ from part (f). Select a set of initial conditions, and be careful to stop the integration when the bar collides with the horizontal surface.

Chapter 10 Notes.

11   Energy Analysis

An Introduction. Investigating the energy of a system can be revealing. For example, without formally solving the governing equations, one can determine if the total energy is constant or changing. This can be useful in understanding plots of the motion trajectories, or can help one check complex numerical simulations, or be used to help decide what control actions should be imparted.

The system energy is a scalar function: there is only one equation. Therefore, generally speaking, an energy analysis cannot be used to completely solve for the motion of a problem. For example, if an aircraft is moving in 3-dimensional space under the action of applied and gravitational forces, then energy alone cannot be used to solve for the aircraft motion time history.

So, although energy is an important concept to understand, it is no substitute for $f = ma$.

Kinetic Energy of a Point Mass. The kinetic energy for a point mass model is a scalar function that depends on the translational velocity.

$$T = \frac{1}{2}m\boldsymbol{v}\cdot\boldsymbol{v} = \frac{1}{2}mv^2 \tag{208.1}$$

Here, v means the magnitude of the velocity vector. Notice that the kinetic energy function is always non negative, $T \geq 0$.

The change in the kinetic energy is due to the externally applied forces. This is best demonstrated by computing the time derivative of T and using Newton's second law of motion.

$$\dot{T} = m\boldsymbol{a}\cdot\boldsymbol{v} = \boldsymbol{f}\cdot\boldsymbol{v} \quad \text{where } \boldsymbol{f} \text{ is the sum of the external forces.} \tag{208.2}$$

This shows that the kinetic energy function is constant if the sum of the externally applied forces is zero or if the sum of the externally applied forces is always perpendicular to the velocity vector.

$$\dot{T} = 0 \quad \text{if } \boldsymbol{f} = 0 \text{ or } \boldsymbol{f} \perp \boldsymbol{v} \tag{208.3}$$

★ 57 Kinetic Energy of a Sliding Block.

A block slides on a horizontal plane. Analyze if the kinetic energy is constant for the cases $M_k = 0$ and $M_k \neq 0$.

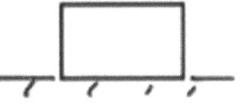

steps 1,2,3,4. (System, FBD, ref frames, vector rep.)

$$\underline{F} = (N - W)\,\hat{n}_2$$

step 5.   kinematics.

$$\underline{v} = v\,\hat{n}_1$$

Compute $dT/dt = \dot{T}$

$$\dot{T} = \underline{F} \cdot \underline{v} = 0 \qquad \text{so kinetic energy is constant.}$$

steps 1,2,3,4. (System, FBD, ref frames, vector rep.)

$$\underline{F} = (N - W)\,\hat{n}_2 - f\,\hat{n}_1$$

step 5.   kinematics.

$$\underline{v} = v\,\hat{n}_1$$

$$\dot{T} = \underline{F} \cdot \underline{v} = -f\,v \neq 0 \qquad \text{so kinetic energy not constant.}$$

Potential Forces. Some forces that act on a system are special in that they are directly related to a scalar function called a *potential function*. Not surprisingly these forces are called potential forces.

The relationship between the potential function and the associated potential force is that the force is related to the gradient of the function. Consider a potential function ϕ that is in terms of Cartesian coordinates, $\phi = \phi(x, y, z)$, where x is the coordinate along a unit vector $\hat{n}_1$, y is the coordinate along a unit vector $\hat{n}_2$, and z is the coordinate along a unit vector $\hat{n}_3$. The gradient of ϕ is computed in the following way.

$$\boldsymbol{\nabla}\phi = \frac{\partial\phi}{\partial x}\hat{n}_1 + \frac{\partial\phi}{\partial y}\hat{n}_2 + \frac{\partial\phi}{\partial z}\hat{n}_3 \tag{210.1}$$

The potential force related to ϕ is $\boldsymbol{f}_\phi = -\boldsymbol{\nabla}\phi$. The subscript ϕ is just to remind us that the force is related to the potential function. The force is suitably called a potential force.

Perhaps the most famous example of all this is the gravitational potential function that gives rise to a constant gravitational force. Consider a two-dimensional setting where y is the coordinate along the $\hat{n}_2$ direction, which points upwards.

$$\phi = mgy \quad ; \quad \boldsymbol{f}_\phi = -\boldsymbol{\nabla}\phi = -mg\hat{n}_2 \tag{210.2}$$

Another common example is the force that a linear spring exerts on a connected body. The potential function is related to the spring elongation or compression along the length of the spring. Consider a one-dimensional setting where x is the coordinate along the $\hat{n}_1$ direction.

$$\phi = \frac{1}{2}kx^2 \quad ; \quad \boldsymbol{f}_\phi = -\boldsymbol{\nabla}\phi = -kx\hat{n}_1 \tag{210.3}$$

The parameter k is the spring constant, and x is the amount of elongation or compression. If $x > 0$, that is, if the spring is stretched along the positive $\hat{n}_1$ direction, then the spring force is in the negative $\hat{n}_1$ direction; if $x < 0$, that is, if the spring is compressed along the positive $\hat{n}_1$ direction, then the spring force is in the positive $\hat{n}_1$ direction.

Total Energy. Potential functions that give rise to potential forces are commonly called potential energies. So, $\phi_{\text{grav}} = mgy$ is the gravitational potential energy and $\phi_{\text{spring}} = \frac{1}{2}kx^2$ is the potential energy of a linear spring.

The total energy of a system is the sum of the kinetic energy and the potential energies.

$$E = T + \phi_{\text{grav}} + \phi_{\text{spring}} + \cdots \tag{211.1}$$

Earlier we studied the change in the kinetic energy by computing the time derivative of T. In a similar way, it is instructive to study the change in the total energy of a point mass by computing the time derivative of E.

$$\dot{E} = \dot{T} + \dot{\phi} = m\boldsymbol{a}\cdot\boldsymbol{v} + \boldsymbol{\nabla}\phi\cdot\boldsymbol{v} = (\boldsymbol{f}_\phi + \boldsymbol{f}_{np})\cdot\boldsymbol{v} - \boldsymbol{f}_\phi\cdot\boldsymbol{v} = \boldsymbol{f}_{np}\cdot\boldsymbol{v} \tag{211.2}$$

This is an important result. First note that the forces $\boldsymbol{f}$ were partitioned into potential $\boldsymbol{f}_\phi$ and nonpotential $\boldsymbol{f}_{np}$ forces. The nonpotential forces are the externally applied forces that cannot be computed from a potential function. Examples include reaction forces from contact (i.e., normal forces) and friction forces. Equation (211.2) shows that the energy function is constant if the externally applied nonpotential forces are zero or if the externally applied nonpotential forces are always perpendicular to the velocity vector.

$$\dot{E} = 0 \quad \text{if } \boldsymbol{f}_{np} = 0 \text{ or } \boldsymbol{f}_{np} \perp \boldsymbol{v} \tag{211.3}$$

⋆ 58 Total Energy of an Inclined Block.

A block slides on a ramp. Friction
acts at the contact surface.
Identify the potential and nonpotential forces. Is energy E
constant?

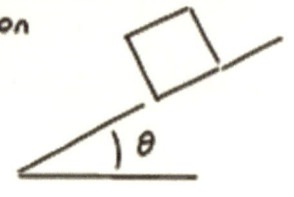

steps 1,2,3,4. (System, FBD, ref frames, vector rep.)

$$\underline{F} = f \,\hat{n}_1 + N \,\hat{n}_2$$
$$- W \cos\theta \,\hat{n}_2 - W \sin\theta \,\hat{n}_1$$

step 5. kinematics
$$\underline{v} = - v \,\hat{n}_1$$

Potential forces. W

Non potential forces. f and N

Compute $\dot{E}$

$$\dot{E} = \underline{f}_{non\,pot} \cdot \underline{v} = (f \,\hat{n}_1 + N \,\hat{n}_2) \cdot (-v \,\hat{n}_1)$$
$$= - f v \neq 0$$

E not conserved.

Energy of a Rigid Body. A rigid body has two types of kinetic energy. There is kinetic energy due to translational motion of the mass center and kinetic energy due to rotational motion about the mass center.

$$T = T_t + T_r = \frac{1}{2}m\boldsymbol{v}_c \cdot \boldsymbol{v}_c + \frac{1}{2}I_c\omega^2 \qquad (213.1)$$

The total energy of a rigid body is the sum of the total kinetic energy and total potential energies, if they are present.

$$E = T_t + T_r + \phi_{\text{grav}} + \phi_{\text{spring}} + \ldots = T + \phi \qquad (213.2)$$

The gravitational potential energy of a point mass model is related to the vertical displacement from a reference line of the point mass. For a rigid body, the gravitational potential energy ϕ_{grav} is related to the vertical displacement of the mass center, and the associated gravitational force acts through the mass center.

⋆ 59 The Compound Pendulum.

A disk and bar are welded together. They swing about the frictionless pivot O. Compute the energy expression for an arbitrary angle θ and speed $\dot{\theta}$.

Bar: $\underline{v}_c = {}^{L}/_{2}\,\dot{\theta}\,\hat{e}_2$ $\qquad \omega = \dot{\theta}$

$$T_{Bar} = T_{transl} + T_{rot}$$

$$T_{Bar} = \frac{1}{2}\,m\,\frac{L^2}{4}\,\dot{\theta}^2 +$$

$$\frac{1}{2}\left(\frac{1}{12}m L^2\right)\dot{\theta}^2$$

$$\phi_{Bar} = -\,mg\,\frac{L}{2}\cos\theta$$

Disk: $\underline{v}_c = L\dot{\theta}\,\hat{e}_2$ $\qquad \omega = \dot{\theta}$

$$T_{disk} = \frac{1}{2}\,M\,L^2\dot{\theta}^2 +$$

$$\frac{1}{2}\left(\frac{1}{2}Mr^2\right)\dot{\theta}^2$$

$$\phi_{disk} = -\,Mg\,L\cos\theta$$

$$E = \frac{1}{2}\,\dot{\theta}^2\left\{\frac{mL^2}{3} + M\left(L^2 + \frac{r^2}{2}\right)\right\} - gL\cos\theta\left(M + \frac{m}{2}\right)$$

The Change in Energy of a Rigid Body. Earlier we studied the change in energy of a point mass model by computing the time derivative of E. In a similar way, it is instructive to study the change in energy of a rigid body model. For simplicity, only a gravitational potential energy is included, and we suppose that the rigid body is subject to forces only: there are no pure moments acting.

$$\dot{E} = \dot{T} + \dot{\phi} = m\boldsymbol{a}_c \cdot \boldsymbol{v}_c + I_c \alpha \omega + \boldsymbol{\nabla}\phi \cdot \boldsymbol{v}_c = (\boldsymbol{f}_\phi + \boldsymbol{f}_{np}) \cdot \boldsymbol{v}_c + \tau\omega - \boldsymbol{f}_\phi \cdot \boldsymbol{v}_c$$

$$= \boldsymbol{f}_{np} \cdot \boldsymbol{v}_c + \tau\omega \tag{215.1}$$

The moments τ are due to the nonpotential forces because, by acting at a point $\diamond$ on a rigid body, the nonpotential forces may create a moment about the mass center.

$$\tau = \boldsymbol{r}_{\diamond/c} \times \boldsymbol{f}_{np} \tag{215.2}$$

Note also that the velocity of the mass center can be related to the velocity of point $\diamond$ on the rigid body.

$$\boldsymbol{v}_c = \boldsymbol{v}_\diamond + \boldsymbol{\omega} \times \boldsymbol{r}_{c/\diamond} \tag{215.3}$$

The vector $\boldsymbol{r}_{\diamond/c}$ is a vector from c to $\diamond$ whereas $\boldsymbol{r}_{c/\diamond}$ is a vector from $\diamond$ to c. Clearly $\boldsymbol{r}_{c/\diamond} = -\boldsymbol{r}_{\diamond/c}$. Equations (215.3) and (215.2) can be used to reduce eq. (215.1) and show that it's the nonpotential forces and the velocity of the rigid body point where the force is applied that's important.

$$\dot{E} = \boldsymbol{f}_{np} \cdot \boldsymbol{v}_\diamond \tag{215.4}$$

If pure moments (designated as τ^*) are acting on a rigid body, then eq. (215.1) still holds true except that τ will contain the pure moments also, $\tau = \boldsymbol{r}_{\diamond/c} \times \boldsymbol{f}_{np} + \tau^*$. Equation eq. (215.4) slightly changes, however, because of the pure moments, $\dot{E} = \boldsymbol{f}_{np} \cdot \boldsymbol{v}_\diamond + \tau^*\omega$.

⋆ 60 Martian Rover Deployment.

A planned Martian rover will roll off of the ramp of a landcraft. The ramp is modeled as a parabolic curve and the rover is

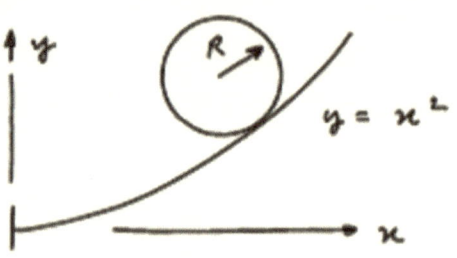

$y = x^2$

modeled as a hoop. The hoop begins from rest at $x = 0.5$ m and rolls without slip down the ramp. Some parameter values are $I_c = mR^2$; $\mu_s = 4/5$; $\mu_k = 1/4$; $m = 20$ kg ; $R = 1/2$ m. Gravity acts downward. Let $g = 4$ m/s². Is energy conserved for the nonslipping motion? compute the translational velocity of the hoop when $x = 0$.

steps 1,2,3,4. (System, FBD, ref frames, vector rep.)

$$\underline{f} = f\,\hat{e}_1 \qquad \underline{N} = N\,\hat{e}_2$$

$$\underline{w} = -mg\cos\theta\,\hat{e}_2 - mg\sin\theta\,\hat{e}_1$$

Non pot. : $\quad \underline{f} \ \& \ \underline{N}$

Pot. : $\quad \underline{w}$

$$\dot{E} = (f\,\hat{e}_1 \cdot \underline{0}) + (N\,\hat{e}_2 \cdot \underline{0}) + 0\,(\dot{\theta}\,\hat{e}_3)$$

$$= 0 \quad \text{so yes, } E \text{ conserved.}$$

Compute energy. $E = T + \phi$. Since E is constant, $E_1 = E_2$

$$T_1' + \phi_1 = T_2 + \phi_2'$$
$$\cancel{0} \qquad\qquad\qquad \cancel{0}$$

$\tan\theta = \dfrac{dy}{dx} = 2x$

so

$\theta = 45°$

$y = x^2$

$$\phi_1 = mgH = mg(y + h)$$
$$= mg(x^2 + R\cos\theta - R)$$
$$= mg\left(\tfrac{1}{4} + \tfrac{1}{2}\tfrac{1}{\sqrt{2}} - \tfrac{1}{2}\right)$$
$$= \frac{mg}{4}(\sqrt{2} - 1)$$

$$T_2 = \tfrac{1}{2}m\,\underline{v_c}\cdot\underline{v_c} + \tfrac{1}{2}I_c\,\omega^2$$
$$= \tfrac{1}{2}m v^2 + \tfrac{1}{2}mR^2\omega^2$$

Roll w/out slip so $v = R\omega$, so $T_2 = mv^2$

so $\phi_1 = T_2$ gives $\dfrac{mg}{4}(\sqrt{2}-1) = mv^2$

$$v = \sqrt{\frac{\sqrt{2}-1}{}}$$

Chapter 11 Problem Set.

1. (M&K 3.174) AggieSatX is in an elliptical orbit around the Earth. The orbit perigee is marked A and at this point the satellite, which is modeled as a point mass, has a velocity of 25000 km/hr with an altitude of 2200 km. (a) Compute the satellite velocity at point B where the altitude is 2500 km. Assume the Earth is a sphere with radius 6371 km. Let $g = 9.825$ m/s^2 be the gravity constant at the Earth's surface. Answer: $v_B = 24170$ km/hr.

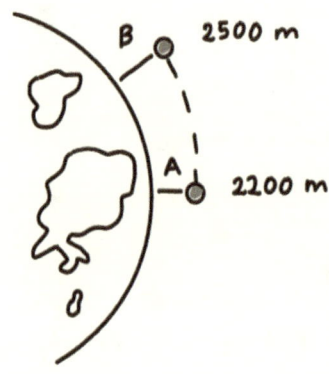

B
2500 m

A
2200 m

Chapter 11 Notes.

12 Docking, Despin, & Thrust

An Introduction. Spacecraft docking and rocket thrust are different sides of the same coin. Satellite despin is somewhere in the middle. A system gains mass in spacecraft docking; a system loses mass in rocket thrust; and mass is constant but redistributed in satellite despin. In this section we take elementary looks at these important aerospace maneuvers.

A Momentum Analysis for Docking. Newton's second law of motion, in either its differential or integral form, applies to a constant mass system of point masses. Here we'll consider the integral form for the special case of rectilinear motion.

$$\int_{t_0}^{t} f \, dt = (m_1 v_1 + m_2 v_2 + \ldots)_t - (m_1 v_1 + m_2 v_2 + \ldots)_{t_0} \tag{222.1}$$

To begin, consider two constant mass systems traveling along a fixed direction such that the trailing mass m_1 overtakes the lead mass m_2. Mass m_1 travels with initial speed $v_1 > 0$ and mass m_2 travels with initial speed $v_2 \geq 0$. In our case, the two systems eventually dock, so $v_1 > v_2$. After docking, m_1 and m_2 travel together with a common velocity $v_1 + \Delta v$. Suppose the docking maneuver occurs over the time interval Δt during which the external forces f are constant.

$$f \Delta t = (m_1 v_1 + m_2 v_1 + m_1 \Delta v + m_2 \Delta v) - (m_1 v_1 + m_2 v_2) \tag{222.2}$$

This expression is some times rewritten using the velocity of m_1 relative to m_2, $u = v_1 - v_2$.

$$f \Delta t = m_1 \, \Delta v + m_2 \, u + m_2 \Delta v \tag{222.3}$$

If enough information is known, then this one equation can provide Δv to determine the final velocity of the docked system.

⋆ 61 Command and Lunar Module Docking.

A command module is positioned to dock with a lunar landing module. There are no significant external forces to consider during the maneuver. Compute the final velocity of the docked system. Consider $v_2 = 0$.

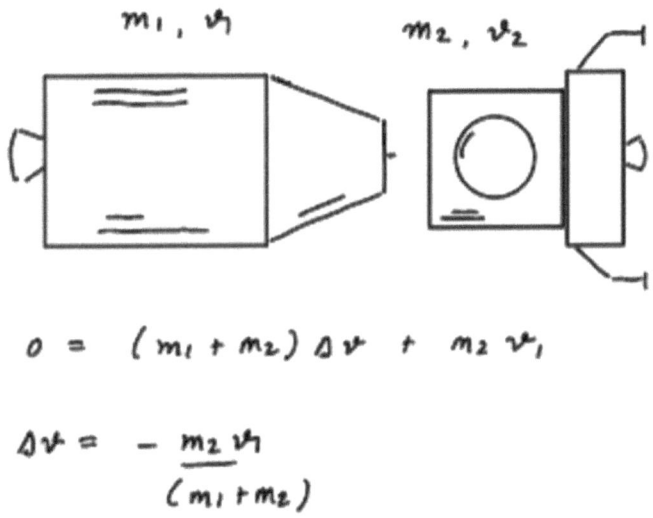

$$0 = (m_1 + m_2) \Delta v + m_2 v_1$$

$$\Delta v = - \frac{m_2 v_1}{(m_1 + m_2)}$$

$$v' = v_1 + \Delta v = v_1 - \frac{m_2 v_1}{(m_1 + m_2)} = \frac{m_1 v_1}{(m_1 + m_2)}$$

Energy Lost in Docking. The kinetic energy of the system composed of two point masses can be computed before and after the docking maneuver.

$$T_{\text{before}} = \frac{1}{2}m_1v_1^2 + \frac{1}{2}m_2v_2^2 \tag{224.1}$$

$$T_{\text{after}} = \frac{1}{2}(m_1 + m_2)(v_1 + \Delta v)^2 \tag{224.2}$$

The difference in these kinetic energy values shows that energy is lost in the docking maneuver. The difference depends on the mass of the vehicles and the relative velocity.

$$\Delta T = -\frac{1}{2}\frac{m_1m_2}{m_1 + m_2}u^2 \tag{224.3}$$

Equation (224.3) is for the special case of no external forces, $f = 0$.

Could a nonzero, constant force f act on the system during Δt to offset the loss in the kinetic energy? Certainly. Although there are two mathematical answers only one makes physical sense, and it depends on the initial momentum p_{before} and initial energy T_{before}.

$$f = \frac{\sqrt{2(m_1 + m_2)T_{\text{before}}} - p_{\text{before}}}{\Delta t} \tag{224.4}$$

Satellite Despin. An integral form of Euler's second law of rotational motion relates the rotational impulse to the change in angular momentum.

$$\boldsymbol{\ell}_c = \dot{\boldsymbol{h}}_c \quad \longrightarrow \quad \int_{t_0}^{t} \ell_c \, dt = (I_c \omega)_t - (I_c \omega)_{t_0} \tag{225.1}$$

When investigating the motion of a spinning satellite, it is common to assume that the torques are negligible. Thus, the angular momentum about the mass center is conserved.

$$(I_c \omega)_t = (I_c \omega)_{t_0} \tag{225.2}$$

Additionally, if there are no torques, then the energy of the spinning satellite is conserved.

$$\left(I_c \omega^2\right)_t = \left(I_c \omega^2\right)_{t_0} \tag{225.3}$$

Equations (225.2) and (225.3) can be used to investigate some general satellite maneuvers, including the classic yo-yo despin maneuver. In the next example, we'll focus on a special case of the yo-yo maneuver which investigates the time to completely despin a satellite. Amazingly, the amount of cable needed to despin a satellite is independent of the initial spin rate.

⋆ 62 Yo-yo Despin.

Earlier we studied some kinematics for a yo-yo device. We computed the velocity of a point mass relative to the mass center of the satellite.

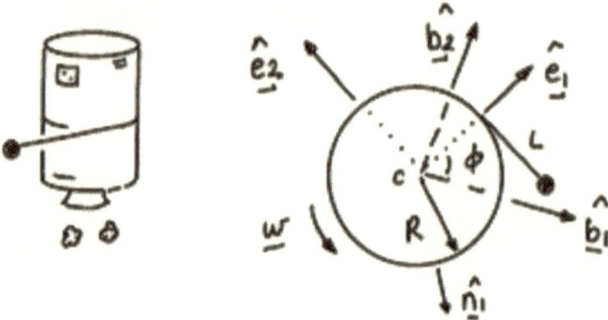

$$\underline{v}\,\text{mass}/_c = R\phi\,(w + \dot\phi)\,\hat{\underline{e}_1} + Rw\,\hat{\underline{e}_2}$$

Assuming that the satellite is only experiencing planar rotational motion, the kinetic energy & angular momentum of the system about C can be easily determined.

$$T = \tfrac{1}{2}\,I_c\,w^2 + \tfrac{1}{2}\,m\left(R^2\phi^2\,(w + \dot\phi)^2 + R^2 w^2\right)$$

$$= \tfrac{1}{2}\,(I_c + mR^2)\,w^2 + \tfrac{1}{2}\,m R^2\phi^2\,(w + \dot\phi)^2$$

$$\underline{h} = I_c\,w\,\hat{\underline{e}_3} + (R\,\hat{\underline{e}_1} - R\phi\,\hat{\underline{e}_2}) \times$$

$$\qquad\qquad m\left(R\phi\,(w + \dot\phi)\,\hat{\underline{e}_1} + Rw\,\hat{\underline{e}_2}\right)$$

$$= \left[(I_c + mR^2)\,w + mR^2\phi^2\,(w + \dot\phi)\right]\hat{\underline{e}_3}$$

These expression hold true throughout the un-winding process, so they hold true in the beginning when $\phi = 0$.

$$T_0 = \tfrac{1}{2}(I_c + mR^2)\,\omega_0^2 \quad ; \quad \underline{h}_0 = (I_c + mR^2)\,\omega_0\,\hat{e}_3$$

If there are no externally applied torques, then

$$T = T_0 \quad \& \quad \|\underline{h}\| = \|\underline{h}_0\| \quad \text{for all time.}$$

Energy:
$$(I_c + mR^2)\,\omega^2 = (I_c + mR^2)\,\omega_0^2$$
$$+ m R^2 \phi^2 (\omega + \dot{\phi})^2$$

Ang. Mom :
$$(I_c + mR^2)\,\omega = (I_c + mR^2)\,\omega_0$$
$$+ m R^2 \phi^2 (\omega + \dot{\phi})$$

Multiplying the angular momentum balance with $(\omega + \dot{\phi})$ and subtracting the kinetic energy balance reveals $\dot{\phi} = \omega_0 \quad \Rightarrow \quad \phi = \omega_0\, t$

This result may be used in the angular momentum balance and one may determine how much cable is needed to completely despin the satellite (i.e., $\omega = 0$)

$$m R^2 \phi_f^2\, \omega_0 = (I_c + mR^2)\,\omega_0 \quad \Rightarrow \quad L_f = R\sqrt{\frac{I_c}{mR^2} + 1}$$

The cable length does not depend on ω_0!

An Instantaneous Form.   Previously, an integral form for rectilinear motion in which one point mass body overtakes another was found.

$$f\Delta t = m_1\,\Delta v + m_2\,u + m_2\Delta v \qquad (228.1)$$

Mass m_1 is the pursuing mass.

Now consider the case $m_2 \ll m_1$ and designate $m_2 = \Delta m$. Furthermore, now consider Δ quantities to be relatively small when compared to their non-Δ counterparts.

$$f\Delta t = m_1\,\Delta v + \Delta m\,u + \Delta m\,\Delta v \qquad (228.2)$$

The final term on the right side is second-order in Δ quantities whereas the other terms are first-order. Dividing the entire equation Δt and passing to the limit that Δ quantities shrink to zero gives a differential form.

$$f = m\dot{v} + \dot{m}u \qquad (228.3)$$

The term $\dot{m}u$ is commonly treated as an external force acting on mass m.

Equation (228.3) can model, for example, an orbiting spacecraft that collects small pieces of orbital debris. In such a case, m and $\dot{v}$ are the instantaneous mass and acceleration of the spacecraft, $\dot{m} > 0$ is the positive mass rate (i.e., the mass of the spacecraft is increasing), and $u > 0$ is the velocity of the spacecraft relative to the orbital debris. Thus $\dot{m}u > 0$, and when it is put on the force side of the equation it appears, or can be interpreted, as a drag or retarding force on the spacecraft.

Equation (228.3) equally models the case of a rocket expelling mass to provide rocket thrust. In such a case, m and $\dot{v}$ are the instantaneous mass and acceleration of the rocket, $\dot{m} < 0$ is the negative mass rate (i.e., the mass of the rocket is decreasing), and $u > 0$ is the velocity of the rocket relative to the exhaust. Thus $\dot{m}u < 0$, and when it is put on the force side of the equation it appears, or can be interpreted, as a thrust force on the rocket.

The Sounding Rocket. The velocity of a rocket that expels mass to provide thrust is governed by an ordinary differential equation.

$$m\dot{v} = f - \dot{m}u \tag{229.1}$$

The term $-\dot{m}u$ is positive and is commonly called the thrust acting on mass m. A traditional case to study is that in which f equals the instantaneous weight $-mg$.

$$m\dot{v} = -mg - \dot{m}u \tag{229.2}$$

This equation can be integrated in a straightforward way when one assumes that u, which reflects the exhaust velocity, and g are constants.

$$v = -gt + u\ln\frac{m_0}{m} \tag{229.3}$$

Here we have taken as limits of integration the particular case that the rocket begins from rest at time $t = 0$.

Further analysis is possible of one assumes the mass burn rate is constant.

$$m = \dot{m}t + m_0 \quad \rightarrow \quad v = -\frac{g}{\dot{m}}(m - m_0) + u\ln\frac{m_0}{m} \tag{229.4}$$

The condition when the fuel has been spent is known as burnout. The velocity at burnout is the maximum that can be attained, and it depends on the exhaust velocity, the mass of fuel in the rocket, and the mass burn rate.

$$v_b = -\frac{g}{\dot{m}}(m_b - m_0) + u\ln\frac{m_0}{m_b} \tag{229.5}$$

Instead of the velocity of the rocket relative to the exhaust and mass burn rate, u and $\dot{m}$, this equation is more commonly written in terms of two engine-performance ratios, the *specific impulse* $I = u/g$ and *thrust ratio* $R = T/(m_0 g) = a_0/g + 1$, where a_0 is the initial rocket acceleration, and a mass ratio $m_* = m_b/m_0$.

$$v_b = Ig\left(\ln\frac{1}{m_*} - \frac{1}{R}(1 - m_*)\right) \quad ; \quad R > 1 > m_* > 0 \tag{229.6}$$

Single Stage to Orbit. This plot shows graphs of burnout velocity v_b versus mass ratio $m_* = m_b/m_0$ over a range of specific impulses $I = u/g$ for a specific thrust ratio $R = a_0/g+1$. The governing equation was presented on the previous page and applies to a single-stage rocket.

$$v_b = Ig\left(\ln\frac{1}{m_*} - \frac{1}{R}(1 - m_*)\right) \qquad (230.1)$$

It is difficult to obtain m_* values less than 0.1 and I values greater than 350, which means burnout velocities beyond 4 mi/sec are basically out of reach. This plot illustrates that single-stage rockets are inadequate for placing satellites into low Earth orbit (altitudes in the range 100 to 1240 mi) because that would require burnout velocities between 4.3 and 4.8 mi/sec.

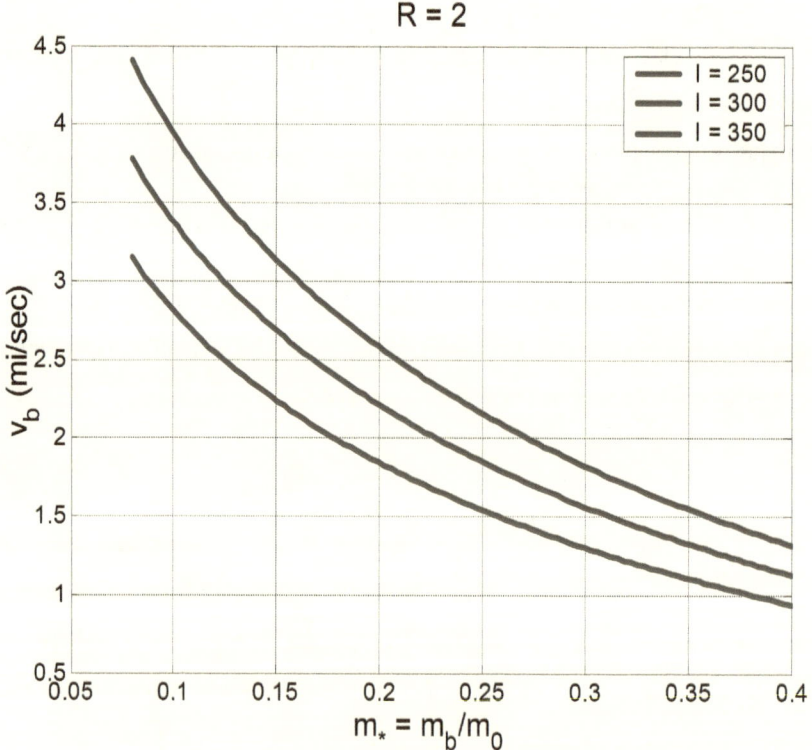

(⋆ **63 Variable Mass Rocket Exercises.**)

Clearly define any new parameters, dimensions, coordinates, variables, etc. that are needed to complete each problem.

1. Develop an equation to compute the time for a single-stage rocket to reach its maximum altitude if the rocket dumps the mass of the fuel container at burnout.

2. Develop an equation for the burnout velocity a two-stage rocket if the rocket dumps the mass of the first fuel container at the first burnout.

(**Maximum Altitude.**)   Again, an expression for the velocity of a single-stage rocket during boost was found to depend on the constant exhaust velocity, the instantaneous mass of the rocket, and the constant mass burn rate.

$$v = -\frac{g}{\dot{m}}(m - m_0) + u \ln \frac{m_0}{m} \tag{232.1}$$

Integrating this equation leads to the altitude during boost.

$$h = -\frac{g}{2\dot{m}^2}(m - m_0)^2 + \frac{u}{\dot{m}}\left(m - m_0 + m \ln \frac{m_0}{m}\right) \tag{232.2}$$

These kinematic expressions can be evaluated at burnout and written using the specific impulse $I = u/g$, the thrust ratio $R = a_0/g + 1$, where a_0 is the initial rocket acceleration, and the mass ratio $m_* = m_b/m_0$. Note that $R > 1 > m_* > 0$.

$$v_b = Ig\left(\ln \frac{1}{m_*} - \frac{1}{R}(1 - m_*)\right) \tag{232.3}$$

$$h_b = -\frac{gI^2}{2R^2}(1 - m_*)^2 + \frac{I^2}{gR}\left(1 - m_* - m_* \ln \frac{1}{m_*}\right) \tag{232.4}$$

After burnout, under the current idealizations, the nonpotential forces cease to act on the rocket. Thus, the energy of the motion is conserved so that the loss in kinetic energy is simply gained by the potential energy, $E = m_b v^2/2 + m_b g h$. The energy constant can be determined by evaluating the energy expression at burnout. Consequently, at any moment during the climb after burnout the following balance equation holds.

$$\frac{1}{2}v^2 + gh = \frac{1}{2}v_b^2 + gh_b \tag{232.5}$$

The maximum altitude is achieved when $v = 0$.

$$h_{\max} = \frac{gI^2}{2}\left(\ln \frac{1}{m_*} - \frac{1 - m_*}{R}\right)^2 + \frac{I^2}{gR}\left(1 - m_* - m_* \ln \frac{1}{m_*}\right) - \frac{gI^2(1 - m_*)^2}{R} \tag{232.6}$$

The first two terms are always positive whereas the last term is always negative. This implies that $h_{\max}$ can be maximized by maximizing the thrust ratio R, which in turn means maximizing the initial acceleration. This choice means quickly burning all of the fuel while leaving the launch pad.

Duct Propulsion.[9] Suppose a vehicle is immersed in a liquid or gas. Suppose the liquid or gas is ducted through the vehicle and expelled at a higher velocity due to thermal or mechanical work. This process provides a means of propulsion for the vehicle, which is commonly called ducted propulsion. We can derive the governing equation of motion for the vehicle using the techniques that we used before.

The differential form for a system that simultaneously gains and loses mass has a simple form.

$$f = m\dot{v} + \dot{m}_{in}u_{in} + \dot{m}_{out}u_{out} \tag{233.1}$$

Here, $\dot{m}_{in} > 0$ whereas $\dot{m}_{out} < 0$. Also, the speeds $u_{in}, u_{out} > 0$ are the relative velocities between the instantaneous system mass m and the intake and exhaust fluids, respectively. Now, suppose the intake fluid is at rest, then $u_{in} = v - v_{in} = v$.

$$f = m\dot{v} + \dot{m}_{in}v + \dot{m}_{out}u_{out} \tag{233.2}$$

Furthermore, suppose the exiting mass rate consists of the intake mass rate and the added fuel rate from the vehicle, $-\dot{m}_{out} = \dot{m}_{in} - \dot{m}$, where $\dot{m} < 0$.

$$f = m\dot{v} + \dot{m}_{in}v - (\dot{m}_{in} - \dot{m})u_{out}$$

or rearranging and omitting the 'out' subscript,

$$m\dot{v} = f + \dot{m}_{in}(u - v) - \dot{m}u \tag{233.3}$$

Equation (233.3) models the rectilinear motion of a vehicle with ducted propulsion, where m is the instantaneous mass, v is the instantaneous speed, f are externally applied forces, $\dot{m}_{in} > 0$ is the mass rate of the entering fluid, $u > 0$ is the velocity of the vehicle relative to the exiting fluid, and $\dot{m} < 0$ is the mass rate of the vehicle (i.e., the mass of the vehicle is decreasing). The *thrust* provided to the vehicle is represented by $T = \dot{m}_{in}(u - v) - \dot{m}u > 0$.

---

[9]See Meriam 1952, p. 567 ff.

⋆ 64 A Jet Airplane.

(Meriam #1090). A jet airplane flies horizontally with constant speed 600 mi/hr = 880 ft/sec. The jet engine consumes air at the rate 140 lb/sec. and uses fuel at the rate 1.30 lb/sec. The exhaust gases have a relative speed of 1800 ft/sec. Determine the drag (resistance forces).

$$\dot{m}_{in} = 140 \frac{lb}{sec} \div 32.2 \frac{ft}{sec^2} \quad ; \quad \vartheta = 880 \frac{ft}{sec}$$

$$\dot{m} = -1.30 \frac{lb}{sec} \div 32.2 \frac{ft}{sec^2} \quad ; \quad u = 1800 \frac{ft}{sec}$$

Along $\hat{n}_1$ direction:

$$m\dot{\vartheta} = f + \dot{m}_{in}(u-\vartheta) - \dot{m}u$$

or

$$m\dot{\vartheta} = -D + T$$

where $T = \dot{m}_{in}(u-\vartheta) - \dot{m}u$

$$T = 4073 \ lb$$

But $\dot{\vartheta} = 0$ so $D = T = 4073 \ lb$

Chapter 12 Problem Set.

1. Recall the page titled *Energy Lost in Docking*. Begin with the first two equations on that page to derive the third.

2. Re-derive the mathematics and expressions pertaining to the *Yo-yo Despin* example.

3. Recreate the mathematics and completely understand all expressions on the page titled *The Sounding Rocket*.

4. A sounding rocket begins from rest and experiences vertical flight in a constant gravity field. Let the relative velocity u be constant. After burnout, the rocket continues to climb until it reaches a maximum altitude. Develop an equation to compute the time for this single-stage rocket to reach its maximum altitude. Clearly define any new parameters, dimensions, coordinates, variables, etc. that are needed to complete each problem. Answer: $t_{\text{total}} = I \ln(m_0/m_b)$.

5. Consider the vertical motion of a sounding rocket in a constant gravity field. Let the relative velocity u be constant. Determine an expression for the mass m as a function of constants and time if it is desired that the vertical acceleration is constant. You may ignore drag. Answer: $m = m_0 \exp(-(a + g)t/u)$.

6. Recall the page titled *Maximum Altitude*. Integrate the first equation to obtain the second. You may use computational software.

7. The 3rd and 4th stages of a rocket are coasting along a fixed direction in deep space with a velocity of 15000 km/hr when suddenly the 4th stage engine ignites causing separation under a constant thrust T and its reaction. At the end of the half-second separation, the 4th stage velocity v_4 is 10 m/s greater than the 3rd stage velocity v_3. The 3rd and 4th stage mass values are assumed to be constant during the separation. The 3rd stage mass is 30 kg and the 4th stage mass is 50 kg. (a) Determine the constant thrust T. (b) Determine the change in kinetic energy of the system before and after separation. (c) Determine the change in translational momentum of the system before and after separation. Answer: $T = 375$ N.

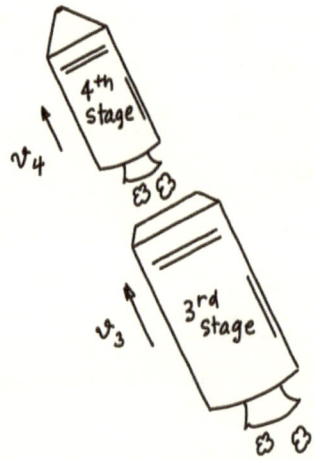

8. A rocket of mass m is coasting along a fixed direction in deep space with a velocity of v_0 when the rocket separates into equal pieces (A & B) due to an internal engine ignition of constant thrust T_1 and its reaction. The separation takes Δt seconds. At the end of separation, the velocity of stage A is Δv greater than stage B. Stage A and B mass values are assumed to be constant during the separation. (a) Determine the constant thrust T_1 and the final velocity of stage A. Your answers should be in terms of m, v_0, Δt, and Δv.

Suppose something similar happens to stage A. That is, stage A splits into equal pieces (C & D) due to an internal constant thrust T_2 and its reaction. The separation takes Δt seconds. At the end of separation, the velocity of stage C is Δv greater than stage D. (Note: Δt and Δv are the same as above.) (b) Determine the constant thrust T_2 and the final velocity of stage C. Your answers should be in terms of m, v_0, Δt, and Δv.

Suppose the process continues ad infinitum. Determine the nth constant thrust T_n and the final velocity of nth remaining stage. Express your answers in terms of m, v_0, Δt, Δv, and n. Answers: $T_n = (1/2^{n+1})\, m\, (\Delta v/\Delta t)$; $v'_n = v_0 + n\,\Delta v/2$.

9. A mischievous boy and his tiger are planning to launch a single stage, homemade model rocket in a horizontal direction. The launch stand is 2 meters tall and the rocket is aimed at Principal Spittle's office wall. The rocket begins from rest, gravity acts downward, and the terrain is flat. Clearly, the rocket will have horizontal and vertical velocity components after launch. Assume the relative velocity u and mass burn rate $\dot{m}$ are constant. Also, assume the mass ratio $m_0/m_b = e = 2.7183\ldots$ (a) Compute the velocity vector at burnout in terms of gravity, specific impulse, thrust ratio, and the mass ratio. Answer: $v_{\text{burnout}} = gI\hat{n}_1 - gI(e-1)/(Re)\hat{n}_2$.

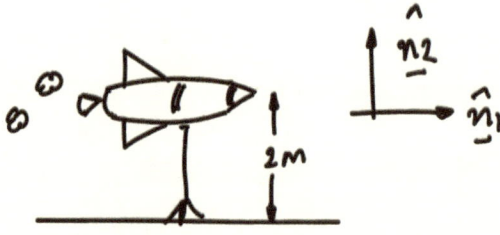

10. A system is composed of a rigid hoop and air-filled balloon that fills the cavity. The hoop radius is R and the system has instantaneous mass m. The system begins from rest. The hoop may roll across a flat surface, and friction acts between the hoop and surface. Gravity acts downward. The balloon vents through a swiveling valve providing a horizontal thrust force $T\hat{n}_1 = -\dot{m}u\hat{n}_1$ at a distance R above the surface. The mass rate $\dot{m}$ and relative velocity u are known. The instantaneous inertia of the system about the mass center is $I_c = mR^2$ where m is the instantaneous system mass. (a) Fully and clearly develop the governing equations for the system. You should find three scalar equations. Assume that the system rolls without slip. You must show FBDs, reference frames, etc. (b) Solve for the instantaneous translational velocity of the mass center as a function of the instantaneous mass, the initial mass m_0, and the relative velocity u. (c) Determine an expression for the translational velocity of the mass center at burnout. (d) Consider the translational velocity of the mass center after burnout. Does it increase? decrease? or stay the same? Clearly explain and defend your answer in an unambiguous way using mathematics and/or complete sentences.

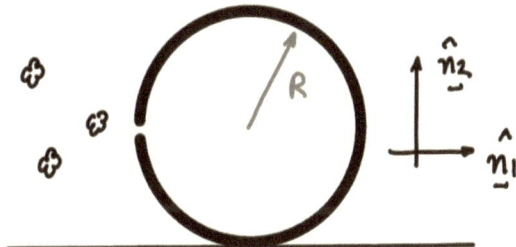

Chapter 12 Notes.

www.ingramcontent.com/pod-product-compliance
Lightning Source LLC
Chambersburg PA
CBHW031121180526
45160CB00005B/42/J